Memory, Place and Identity

This book bridges theoretical gaps that exist between the meta-concepts of memory, place and identity by positioning its lens on the emplaced practices of commemoration and the remembrance of war and conflict.

This book examines how diverse publics in different places relate to their wartime histories through engagement with everyday collective memories. Specifically addressing questions of place-making, displacement and identity, contributions shed new light on the processes of commemoration of war in everyday urban façades and within generations of families and national communities. Contributions seek to clarify how we connect with memories and places of war and conflict. The spatial and narrative manifestations of attempts to contextualise wartime memories of loss, trauma, conflict, victory and suffering are refracted through the roles played by emotion and identity construction in the shaping of post-war remembrances. This book offers a multidisciplinary perspective, with insights from history, memory studies, social psychology, and cultural and urban geography, to contextualise memories of war and their 'use' by national governments, perpetrators, victims and in family histories.

Danielle Drozdzewski is a Human Geographer and Senior Lecturer in the School of Humanities and Languages at the University of New South Wales, Sydney, Australia.

Sarah De Nardi is Research Associate in Cultural Geography at the University of Durham, UK.

Emma Waterton is Associate Professor in the Institute for Culture and Society at University of Western Sydney, Australia.

Routledge Research in Culture, Space and Identity

Series editor: Dr Jon Anderson, School of Planning and Geography, Cardiff University, UK

The *Routledge Research in Culture, Space and Identity Series* offers a forum for original and innovative research within cultural geography and connected fields. Titles within the series are empirically and theoretically informed and explore a range of dynamic and captivating topics. This series provides a forum for cutting-edge research and new theoretical perspectives that reflect the wealth of research currently being undertaken. This series is aimed at upper-level undergraduates, research students and academics, appealing to geographers as well as the broader social sciences, arts and humanities.

Published

Memory, Place and Identity: Commemoration and Remembrance of War and Conflict
Edited by Danielle Drozdzewski, Sarah De Nardi and Emma Waterton

Forthcoming

Surfing Space
Jon Anderson

Trauma and Violence in Place: Cultural and Environmental Wounding
Amanda Kearney

Arts in Place: The Arts, the Urban and Social Practice
Cara Courage

Memory, Place and Identity

Commemoration and remembrance
of war and conflict

**Edited by Danielle Drozdzewski,
Sarah De Nardi and Emma Waterton**

Routledge
Taylor & Francis Group

LONDON AND NEW YORK

First published 2016
by Routledge
2 Park Square, Milton Park, Abingdon, Oxon OX14 4RN

and by Routledge
711 Third Avenue, New York, NY 10017

First issued in paperback 2018

Routledge is an imprint of the Taylor & Francis Group, an informa business

British Library Cataloguing-in-Publication Data
A catalogue record for this book is available from the British Library

Library of Congress Cataloging-in-Publication Data
Names: Drozdzewski, Danielle, editor. | De Nardi, Sarah,
editor. | Waterton, Emma, editor.
Title: Memory, place and identity: commemoration and remembrance of war and conflict/edited by Danielle Drozdzewski, Sarah De Nardi, Emma Waterton.
Description: New York, NY: Routledge, 2016. |
Series: Routledge research in culture, space and identity | Includes bibliographical references and index.
Identifiers: LCCN 2015047071 | ISBN 9781138923218 (hardback) |
ISBN 9781315685168 (e-book)
Subjects: LCSH: War and society. | Collective memory. | War memorials.
Classification: LCC HM554 .M463 2016 | DDC 303.6/6—dc23
LC record available at http://lccn.loc.gov/2015047071

ISBN 13: 978-1-138-54717-9 (pbk)
ISBN 13: 978-1-138-92321-8 (hbk)

Typeset in Times New Roman by
Keystroke, Station Road, Codsall, Wolverhampton

Contents

Figures

Contributors

Maoz Azaryahu is Professor for Cultural Geography at the University of Haifa in Israel. His research focuses on urban and landscape semiotics, on the cultural and historical geographies of public memory in Israel and in Germany, as well as landscapes of popular culture and the cultural history of places and landscapes. His books include *Von Wilhelmplatz to Thälmannplatz: Politische Symbole im öffentlichen Leben der DDR* (1991), *State Cults. Celebrating Independence and Commemorating the Fallen in Israel 1948–1956* (in Hebrew, 1995), *Tel Aviv: Mythography of a City* (2006) and *Namesakes: History and Politics of Street Naming in Israel* (in Hebrew, 2012).

Carolyn Birdsall is Assistant Professor and MA Program Director, Department of Media Studies, University of Amsterdam, the Netherlands. Her monograph *Nazi Soundscapes: Sound, Technology and Urban Space in Germany, 1933–1945*, was published in 2012 (also available through open access: www.oapen.org). Her current project examines broadcast archives in mid-twentieth-century Europe, in view of conflict heritage and the politics of national history and memory cultures.

Sarah De Nardi is Research Associate in Cultural Geography at the University of Durham, UK. Her interdisciplinary research intersects embodied understandings of space and materialities from late Prehistory to the recent past. In particular, her work focuses on the spatial and material heritage of violence and identity in the aftermath of the Second World War in Europe. Her post-doctoral research at the University of Hull spatialised the experience of Second World War Resistance fighters by bringing to the fore the more-than-representational affective heritage of the conflict. At Durham, she is part of an international project investigating the role of museums in the re-enchantment of everyday artefacts, tracing object journeys 'from trash to treasure'. Sarah is about to publish her first monograph, entitled *The Poetics of Conflict Experience: Materiality and Embodiment in Second World War Italy* (forthcoming).

Jason Dittmer is Professor of Political Geography at University College London, UK. His current research is on assemblage theory and its relationship to international relations. He is the author of *Captain America and the Nationalist*

Superhero: Metaphors, Narratives, and Geopolitics (2013) and *Popular Culture, Geopolitics, and Identity* (2010). He is the editor or co-editor of *Geopolitics: an Introductory Reader* (2014), *The Ashgate Research Companion to Media Geography* (2014), *Comic Book Geographies* (2014) and *Mapping the End Times: American Evangelical Geopolitics and Apocalyptic Visions* (2010).

Lia Dong Shimada is a cultural geographer based at the University of Roehampton's Susanna Wesley Foundation, London, UK. She holds a doctorate in Geography from University College London and a Master of Arts degree in Theology and Religious Studies from King's College London. As a conflict practitioner, she has worked with the community peacebuilding process in Northern Ireland and with reforestation movements in Nepal, Ireland and Madagascar. Alongside her academic work, Lia runs a freelance consultancy practice specialising in mediation and organisational diversity.

Danielle Drozdzewski is a Human Geographer and Senior Lecturer in the School of Humanities and Languages at the University of New South Wales, Sydney, Australia. Her research focuses on cultural memories and the interlinkages of these with identity and place. She is interested in how cultural memory is encountered in public spheres and private spaces, between and within generations of families, in homes and with migrants in diaspora. Her current project on the 'Geographies of Everyday Memorialisation' focuses on the influence of mobilities on and at memory sites in post-war landscapes. She has recently published 'Using history in the streetscape to affirm geopolitics of memory' in *Political Geography* (2014) and has edited a special issue in *Emotions, Space and Society* (2015) on researcher trauma, which included her own paper on 'Retrospective reflexivity: the residual and subliminal repercussions of researching war'.

Joshua Hagen is Professor of Geography at Marshall University, West Virginia, USA. He has published widely on issues related to urban planning, historic preservation, and nationalism, including in *Journal of Historical Geography*, *Environment and Planning D: Society and Space*, *Cultural Geographies*, *Political Geography*, *Journal of Urban History*, *Eurasian Geography and Economics*, *Nationalities Papers*, and *Annals of the Association of American Geographers*. He has also co-authored works on geopolitics and border studies, most notably *Borders: A Very Short Introduction* (2012) and *Borderlines and Borderlands: Political Oddities at the Edge of the Nation-State* (2010). He is currently working on two full-length research monographs related to Nazi Germany: *Building Nazi Germany: Space, Place, Architecture, and Ideology* (forthcoming) and *Dictating the Past: Place and Memory in Nazi Germany* (also forthcoming).

Nuala C. Johnson is Reader in Geography at Queen's University Belfast, UK. Her research focuses on nationalism and the politics of identity; public memory and monuments; literary spaces; and the historical geographies of science particularly in relation to botanical gardens and botanical illustration. She authored *Ireland, the Great War and the Geography of Remembrance* (2003) and *Nature Displaced, Nature Displayed: Order and Beauty in Botanical Gardens* (2011), and edited *Culture and Society* (2008), *Companion to Cultural Geography* (2003) and *The Wiley-Blackwell Companion to Cultural Geography* (2013).

Duncan Light works in the Department of Tourism and Hospitality, Bournemouth University, UK. A cultural geographer by background, he has research interests in the relationships between urban landscape, memory and identity in post-socialist Romania. He has published widely on this topic in journals such as *Europe-Asia Studies, Journal of Historical Geography, Transactions of the Institution of British Geographers* and *Annals of the Association of American Geographers*.

Hamzah Muzaini is currently Assistant Professor, Cultural Geography Chair Group, Wageningen University, the Netherlands. From August 2016, he will be Assistant Professor, Southeast Asian Studies Department, National University of Singapore. Before entering academia, he was a heritage consultant, and the curator of the Changi Chapel and Museum, a site dedicated to the remembrance of men and women who were incarcerated by the Japanese in Singapore during the Second World War. His research interests particularly relates to war heritage and memoryscapes, conceptualised around postcolonial theory, materiality and the spatial politics of power and resistance particularly by state and non-state actors in public and everyday spaces. He has also published on backpacking and dark tourism in Southeast Asia, international peace and heritage museums, and Singapore's transborder geographies and histories. His present research looks at 'heritage from below' within the context of cultural theme parks, memory politics as played out through memoryscapes and on spatialities of forgetting and immanence.

Quentin Stevens is Associate Professor and Director of the Centre for Design and Society in the School of Architecture and Design at RMIT University, Melbourne, Australia. He is author of *The Ludic City: Exploring the Potential of Public Spaces* (2007), co-author of *Memorials as Spaces of Engagement: Design, Use and Meaning* (2015) and editor of *Loose Space* (2007), *Creative Milieux* (2015), *The Uses of Art in Public Space* (2014) and *Transforming Urban Waterfronts* (2010).

Shanti Sumartojo is Research Fellow in the School of Media and Communications at RMIT University, Melbourne, Australia. She is the author of *Trafalgar Square and the Narration of Britishness, 1900–2012: Imagining the Nation* (2013) and editor of *Nation, Memory and Great War Commemoration: Mobilizing the Past in Europe, Australia and New Zealand* (2014). Her main research interest is the constitution and experience of atmospheres, and current projects include investigations of First World War commemoration; the 'feel' of aspects of design; and the effect of public art on the experience of public space.

Emma Waterton is Associate Professor based at Western Sydney University, Australia, in the Institute for Culture and Society. Her research explores the interface between heritage, identity, memory and affect. Her most recent project, 'Photos of the Past', is a three-year examination of all four concepts at a range of Australian heritage tourism sites, including Uluṟu Kata-Tjuṯa National Park, Sovereign Hill, the Blue Mountains National Park and Kakadu National Park. She is author of *Politics, Policy and the Discourses of Heritage in Britain* (2010) and co-author of *Heritage, Communities and Archaeology* (with Laurajane Smith, 2009) and *The Semiotics of Heritage Tourism* (with Steve Watson, 2014).

Ross Wilson is Senior Lecturer in Modern History and Public Heritage at the University of Chichester, UK. He has written on the experience, representation and memory of the First World War in Britain and the United States. His wider research focuses on issues of museum, media and heritage representations in the modern era. This work has been published in the books *Representing Enslavement and Abolition in Museums* (2011), *Landscapes of the Western Front* (2012), *Cultural Heritage of the Great War in Britain* (2013) and *New York in the Great War: Making an American City* (2014).

Andrea Witcomb is Professor in Cultural Heritage and Museum Studies at Deakin University, Australia, where she directs the Cultural Heritage Centre for Asia and the Pacific and is the Deputy Director (Governance) of the Alfred Deakin Institute for Citizenship and Globalisation. She has a long-standing interest in the ways in which exhibition practices can be used to create conversations across cultural differences. She has focused on the use of immersive exhibition practices to achieve this end, looking in detail at the poetic side of exhibitions – how objects, first-person narratives, multimedia and sensorial modes of communication are used to produce an affective experience for museum visitors that has the potential to challenge collective memories and understandings. Andrea is the author of *Reimagining the Museum: Beyond the Mausoleum* (2003), *From the Barracks to the Burrup: the National Trust in Western Australia with Kate Gregory* (2010) and co-editor with Chris Healy of *South Pacific Museums: Experiments in Culture* (2006, 2012), as well as contributing many book chapters and journal articles.

Craig Young is Professor in Human Geography in the Division of Geography and Environmental Management at Manchester Metropolitan University, UK. His research interests include a focus on geographies of urban change in post-socialist contexts and the cultural geographies and politics of identity (from the individual to the urban and the nation) in the context of post-socialist transformation, particularly in the former Eastern Europe. Research in Romania has focused on issues of identity, memory, commemoration and the nation as played out through particular urban landscapes and public spaces, especially those surviving from the state-socialist period.

Acknowledgements

The idea for this collection emerged after the editors all presented in the 'Narrating and Remembering Landscapes of Trauma and Conflict' session at the Royal Geographical Society/Institute of British Geographers Conference in London, 2013. While we are each originally trained in differing disciplines – cultural geography, landscape archaeology and heritage studies respectively– our current research interests in memory, and the specific geographical focus on memory and identity in, through and of place, brought this collection to fruition. Overlaying the mutual recognition that memory is most constructively understood when considered as part of a tripartite relationship with identity and place, we were all also working in the borderlands of conflict and war studies. For us, the longevity of memory emanating from war and conflict was overwhelmingly apparent, as was its influence on the (re)production, maintenance and transmission of identity narratives – personal, familial, national and supranational – temporally as well as both inter- and intra-generationally.

We are grateful to all our contributors for their insightful and thought-provoking efforts. We have been delighted not only by the breadth of expertise, but also that the authors in this collection have challenged how we 'do' memory by drawing focus on the body and the non-human components of our memory-making and places. We would like to thanks Faye Leernik at Routledge for her editorial assistance in bringing this collection together. For Danielle and Sarah, this collection is (we hope) the first among many scholarly books; we are both indebted to Emma's sound, sage and always perfectly executed expert advice. The editors also extend sincere thanks to Professor Elaine Stratford for her editorial advice on the proposal and introduction. Sarah is indebted to her colleague Pier Paolo Brescacin and the other staff of ISREV Resistance Studies Institute in Vittorio Veneto, Italy.

We thank the National Gallery of Ireland for permission to reprint William Orpen's Holy Well (1916). We also thank the Imperial War Museum for the permission to reprint the following artwork: William Orpen 'Dead Germans in a Trench' 1918. © IWM (Art.IWM ART 2955); William Orpen 'The Mad Woman of Douai' 1918. © IWM (Art.IWM ART 4671); John Lavery 'Munitions, Newcastle, 1917'. © IWM (Art.IWM ART 1271); and John Lavery 'A Convoy, North Sea, 1918. From N.S. 7'. © IWM (Art.IWM ART 1257). For permission to reprint Lawrence Kirk's 'The Herald of Jericho', we thank the Upper Springfield Development Trust. For permission to reprint 'Divis Mountain', we thank Kris Gillen. We thank the IDF and Defense Establishment Archives for permission to reprint Bab el Wad, 1948, and the Israel Philatelic Service for permission to reprint the Israel postal company Memorial Day Postage Stamp, 2003: 'The armored cars on the road to Jerusalem'.

On a personal note, Danielle wishes to thank William for his support, patience and capacity to continue to listen about the book. To Flynn and Oliver, thank you for understanding (mostly) when mum needed to work. Sarah would like to thank the Second World War veterans and their families who took part in the research upon which this chapter draws. Meeting these brave men and women has been a humbling, unforgettable and personally enriching experience. *Grazie*! Emma is grateful to her family, friends and colleagues for enduring her distracted state during the completion of this book.

1 The significance of memory in the present

Danielle Drozdzewski, Sarah De Nardi and Emma Waterton

Memory is a powerful tool. It invokes the senses: a smell, a familiar touch, an image, a sound once heard before can transport us not only to different times but also to different places (see Tuan 1977). With each memory stimulus we conjure the context of those places, whether they be spatially defined and geographically delimited or immaterial, allegorical places – we situate our memories to place and to time. Memory can connect us to our individual pasts and to the past (as we know it) of those of our closest kin, and does so through the narration of family histories. These connections can be facilitated by sharing ancestral objects, passing down photographs and family names, and reciting habitual practices such as cooking, singing and speaking other languages. Place memory lies at the core of this volume, as illustrated by contributions that acknowledge how deeply this kind of memory is enmeshed in everyday corporeality through 'practices of incorporation' and 'practices of inscription' (Hill 2013: 381). These entanglements link our present-day places and contexts to our pasts and to those of our forebears. Our ability to locate memories shows how the absence of memory is 'evoked, [and] made present, in and through enfolded blendings of the visual, material, haptic, aural, olfactory, emotional-affective and spiritual' (Maddrell 2013: 505). As Bell (1997: 813) has reasoned, places are 'personed . . . even when there is no one there'. Extending Bell's thought to the non-human, places can also be occupied by the presence of objects, animals, thoughts and so on. Thus, we can use place in memory as a positioning tool. The focus on the spatiality of memory is the point of departure for this collection, which comprises the concomitant and geographically contextualised discussions of how these memories are positioned in place and used in the construction and maintenance of identities.

We often think about memory most readily as a personal 'thing'. But memory has a 'sticky' resonance (after Ahmed 2010), drawing in wider contexts and places, including nations and places within them. Memory's adhesive quality attends our capacity to remember trauma. Indeed, as Till (2005: 108) has argued, in 'societies that experienced violence, individuals return to particular places to revisit difficult feelings of loss, grief, guilt and anger'. Likewise, Nordstrom (1997: 4) has reasoned that we revisit these memories, literally and figuratively, and that 'because the encounter with violence is a profoundly personal event, it is fundamentally linked to processes of self-identity and the politics of personhood'.

Such is the conclusion reached by the authors of chapters in this volume, who variously suggest that encounters with violence – whether direct or indirect – affect place and space in meaningful ways.

Violence thus facilitates the production of remembrance-scapes; while such 'scapes' undoubtedly have performative capacities, they simultaneously locate memory's absent presence, too. In respectfully navigating our way through the difficult territories of trauma, we are mindful of the impact that emotion and embodiment of violence and conflict can have on both individuals and society, and its affects – both on those recounting, sharing and (re)visiting it (De Nardi 2015; Drozdzewski 2015). Such mindfulness has opened up spaces in which to consider the longevity of memory in place notwithstanding its traumatic impact (e.g. Saunders 2004; Diken and Bagge Laustsen 2005). These sorts of affective and emotional connections between person and place develop, for example, through the presence of people at a particular commemoration, and in the absence of those being commemorated. Our witnessing of memory's absent-presence, together with the increasing patronage on days sacred to national remembrance, means that places of and in memory cannot be explained away as solely dictated by a linear, mainstream narrative or representational process or material markers of memory.

Physical places that may have been erased by the ravages of war can 'remain present, yet invisible, in the city' and in our memories (Till 2005: 101–102). War wounds cities: it marks them as places that have been 'harmed and structured by particular histories of physical destruction, displacement and individual and social trauma resulting from state-perpetrated violence' (Till 2012: 6). Incidents of war and conflict are remembered for their traumatic impacts, and for 'unsettling ghosts of place' (Bell 1997: 827) that firmly imprint inhumanity, despair, deprivation and struggle – and it is because these imprints are uncomfortable and unsettling that we remember them, individually and collectively. War and conflict codify sets of behaviours as atrocities, as immoral. We memorialise such events to remember victims, and to remember our capacity to act in such ways. In remembering war, we remember victory and suffering – both have been used by ruling elites to convey much about war's influence on the character of nations and of the capacity of peoples to resist, endure and succeed. While war is often experienced and remembered collectively – on a national scale – it is also experienced and remembered by individuals, whose encounters with it may or may not accord with how the ruling elite chooses to commemorate, if it chooses to commemorate at all.

The power of commemorative choice (Nora 1989) is an overarching component of investigations concerning public memorialisation. Memory is 'spatially constituted' whether in 'concrete and physical' form such as monuments and museums, or in 'non-material' form such as narrative, discourse and stories of the war (after Hoelscher and Alderman 2004: 349). With this power also comes the authority to *place* memory, which portends to memory's purposeful positioning in place. In the case of war memorials, memorialisation might occur where an actual battle or event took place. It might also be positioned in a strategic locale in a city – a busy thoroughfare or public square perhaps. As Birth (2006: 182) points out, creating and maintaining monuments often involves a great deal of attention being paid to the discursive messages that accompany sensory impression. Marking a certain memory in place – in a city square, for example – is thus strategic; it asserts a ruling elite's interpretations of the past and its notions of identity in the present. By using plaques, candles, flowers, national narratives of heroism, remembrance days, and by creating reverential atmospheres, those who remember provide the tools for varied publics to read, encounter, feel and experience that event.

Pivotal to memory's power, then, is its politics or intrinsic usefulness (see Said 2000). A politics of memory speaks to the rationales and operations of memory, as well as the motivations behind place-based choices of representation for certain groups of people. Our contention in this collection is that memory of place is crucial to understanding how identities are rooted to places; comprehending this link may help us better understand reasons for war and conflict in the first place. Consider how, for example, sites of war and conflict in the Ukraine, Syria and Iraq demonstrate the entanglement of memories of place and the rootedness of constructions of identity in, and to, those places. A critical geopolitics of memory 'pits the division and marking of space as a contest' (Drozdzewski 2014: 66) in which the complex relations of place/people/identity become materialised in conflict and invariably fuel territorial incursions into places that certain groups hold in their memory, but do not physically occupy or have authority over. There is longevity to memories of territorial subjugation, whether these remembrances are personally and/or collectively recalled, such as in national groups. For example, the First World War centenary (2014–2018) has so far seen nations once allied to the Western Front commemorate the seventieth anniversaries of different battles including the Gallipoli campaign, the Battles of the Somme, Frommel and Ypres, as well as D-Day. Despite direct lines of (familial) lineage to the First World War and the Second World War in rapid decline, war remembrance and the centrality of championing a nation's collective identity (and as a corollary its cohesiveness to war remembrance) continue to gain currency.

It seems fitting, then, that we borrow from Jones (2011: 2) to submit that 'memory makes us what we are'. Memories both inform and are informed by identities and these articulations take different forms in different places. The mobilisation of memory has the capacity to transform places and keep our articulations of places of, and in memory, fluid. We (re)construct memory in our present-day contexts (see Halbwachs [1926] 1992) and these social frameworks of memory have significance for how contemporary understandings of past events colour remembrance of those same events. In a rapidly changing world, where an ever present 'danger' of the Other retains a sustained presence in both political and popular discourse, assertions of what we (including we as the nation) *are* by focusing on what we are *not* remains at the forefront of politicians' and policy-makers' rhetoric. Reacting to this seeming stoicism in how identity is broached, Macpherson (2010: 7) has questioned why 'in a world of potential and movement' have 'patterns of life become so sedimented and static'. Part of the answer to Macpherson's question can surely be found in analyses of how the framing of identity draws from past (and in some cases present) engagements in war and conflict in ways that cement particular interpretations of identity as collective and shared among citizens of a nation. We contend that as scholars we should agitate for a greater recognition of how a politics of memory is used to attain political and territorial advantage. By confronting and delineating how memory is used for ill purpose, we have the capacity to exercise an ethics of care (see, for instance, Cloke 2002; Olson 2015) in our analysis 'of those evils which, in our time, most menace our capacity to form human lives' (Curtis 1999: 12).

While developing a more nuanced understanding of the intersections between place, identity and memories of war holds relevance for expanding our under-standings of the past, it is also 'fundamental to becoming' (Jones 2011: 2) and thus has continued potency for framing future responses to war and conflict. Understanding memory as a process bound to action is also central to the dynamic (see Salerno and Zarankin 2014), vibrant affective life of places of conflict. Far from being static – even when literally 'set in stone' through a monument – cultural memory is 'an activity occurring in the present, in which the past is continually modified and redescribed even as it continues to shape the future' (Bal 1999: vii). While at the forefront of academic discourse in studies of culture and society since the cultural turn there have studies of identity, and particularly of ethno-cultural identities and multiculturalism, yet there has also been relatively little attention focused on understanding how memories and their politicisation play a role in forming and maintaining identities. Indeed, Johnson (1995: 52) declared that geographers have hesitated to address 'the ways which national cultural identity at the popular level is constructed, maintained, or chal-lenged'. This collection addresses this diffidence by examining how memories of war work both as key components in the constructions of individual, familial and national identities and as markers of places.

Considering memory, identity and place

Integral to our focus on the nexus of memory, place and identity has been an attempt to build on authors such as Butler (1993, 1997, 2004), Taylor (2013) and Barad (2007), who have challenged the centrality of representational thinking and discourse as primary modes of data analysis. We are also indebted to Birth (2006: 176), who has contended that 'remembering is far more than the written word ... it can rely on buildings, spaces, monuments, bodies and patterns of representing self and others'. To guide our approach, we have borrowed from Barad's work (2003, 2007) in order to better understand the intersections of place, memory and identity by examining interactions between human and non-human elements in memoryscapes. This strategy, we think, will serve to open further the spaces of enquiry needed to examine both how events happen to intersect and influence ourselves and our lives, and how our thinking, feelings, emotions and affects are influenced and shaped by the agency of humans (our respondents, passers-by, visitors to memorials) and non-human actants (places, objects, atmospheres). Thus, we have followed Macpherson's (2010: 8) call that there is 'need to attend to the agency of things as well as people', and Birth's (2006: 169) proposition that 'as much as humans may seek to mould the material of memory in particular ways, this may not always be successful when the material itself can exert its own influence on humans in unpredictable ways'. Drawing on the rich seams of embodied experience in this collection, we contend that memory and places of memory have the capability to move us and to generate, for example, haptic, somatic and spontaneous responses.

A key agenda for this collection is to understand the linkages among memory, identity and place in the context of war and conflict. To advance how we think about memory, memorialisation and remembrance, we have turned – both theoretically and methodologically – to those ways of thinking most commonly associated with affect theory, the more-than-human, the more-than-representational and the post-human. Notions of practice and experience are paramount. A key seam binding the chapters in the collection together, then, is the use of innovative and adaptive qualitative methodologies that position research as lived process and not simply product. Cumulatively, the chapters shift and expand memory research beyond more traditional and normative conceptions as routinely articulated in and across landscapes and in the cultural calendar, to something that involves fluid, multiple and often unexpected configurations and interpretations that can be variously and perhaps even simultaneously felt, embodied and encountered.

Underpinning this focus are other intellectual labours articulating the convergence of landscape, heritage and spaces of remembrance (see also Johnson 2014; Tolia-Kelly 2004). The work of Dwyer and Alderman (2008) lays out three distinct approaches to studying the nexus between place and memory: first, memorial landscapes as text; second, memorial landscapes as arena; and, third, memorial landscapes as performance. Through this last approach they have sought to recognise the 'important role that bodily enactments, commemorative rituals, and cultural displays occupy in constituting and bringing meaning to memorials, suggesting that the body itself is a site of memory' (Dwyer and Alderman 2008: 166). Similarly, Brockmeier (2002: 8) has drawn attention to the value of a post-positivist approach to memory, as localising 'memory in culture and, as a consequence, understand[ing] remembering as a cultural practice – be it under the name of social, collective or historical remembering'. The contributions to this book advance and develop Dwyer and Alderman's position by considering how memories of war and conflict – along with powerful attempts to override them – manifest physically, spatially and temporally in place.

In situating our discussion, two key technologies of memory-making cut across the chapters: body memory and non-human agency in memory-making.

Body memory

Memory is articulated, felt, enacted and experienced through the body. The body is thus a place of memory, and it also has mobility through places *in* memory. These critical bodily engagements of, and with, memory substantiate a distinct deviation from more traditional approaches that privileged memory in terms of the textual, the visual and the discursive: as what we read, see and think. Such ruminations align with the Cartesian binary of mind/body, where knowledge gained through the body was seen as tainted by the body's physicality and its inability for rationale and reason, distinctive from the mind's cognitive capacity for logical intellect. In framing and aligning this binary as masculinist, Rose (1993: 7) has argued that it 'assumes a knower who believes he can separate himself from his body, emotions, values, and past experiences so that he and his through are autonomous, context-free and objective'. This collection's authors have not disregarded representational forms of memory per se, but instead have considered Waterton's (2013: 67) proposition about how thinking about memory brings together 'cognition with impulse, intuition and habit' by exploring how memory is sensed, how it can be tactile, how is can elicit bodily responses, how it is felt. For Waterton (2013: 67), too, processes of meaning-making occur 'within action and interactions with other people and the world around us'. And as Langer writes (1989: 41), an awareness of the body is 'inseparable form the world of [its] perception . . . I perceive always in reference to my body'. In short, by opening up (and not shutting out) our analyses to the possibilities of encounter, embodiment and experience of the body, more nuanced understandings of places of war and conflict have been possible. Bodies, as Longhurst (2001: 5) has contended are 'always in a state of becoming with places'.

In our collection, we see body memory enacted in two ways. First, the body is moving through and encountering sites of memory at specially designated memory sites, such as the Australian War Memorial (Witcomb; Dittmer and Waterton; Sumartojo and Stevens; Wilson), at memorials positioned in the everyday landscape at the location of specific events (Drozdzewski; Muzaini; Azaryahu; Young and Light), and in places where war and conflict are ongoing (Johnson; Shimada; Hagen). How the body encounters (and reacts to) these places of memory differs; a key variable among these differences is the politics of representation of the memory and the embedding of that narrative within the social frameworks of each individual's notions of identity. For Wilson (Chapter 13), the body is the agent of witnessing, and in light of declining first-hand witnesses to both the First World War and the Second World War, for example, the body holds the moral, social and political obligation to the continuance of wartime commemoration. The positioning of the body in places of, and in, memory, can generate such obligatory responses. Explaining Foucault's theorisation of the body as demonstrative of the disciplining gaze, Valentine (2001: 28) has stated that the body can be seen as being 'constituted within discourse and that different discursive regimes produce different bodies'. In our collection, this production has direct significance for how bodies respond to memorial spaces and are compelled to witness events through their purposeful framing. Witcomb (Chapter 12), Sumartojo and Stevens (Chapter 11), Dittmer and Waterton (Chapter 10) all discuss how our bodies move through nationally significant memorial sites that have been purposefully designed to encourage us to think and feel a certain way about the nation's contribution to war.

Second, the body is also a site situated in places of war and conflict, giving rise to place-based memories of war. Muzaini (2015: 109) has argued that the 'body may be seen as a receptor of "the immanent past" . . . when it is affected by itself, where internal (rather than external) stimuli allow for memory returns'. In our collection, memories of war are enacted and/or prompted through narrative (De Nardi), sound (Birdsall; Sumartojo and Stevens), art (Shimada; Johnson) and the everyday landscape (Muzaini; Drozdzewski; Hagen) – all impelling the body to respond to recollection. As Birdsall (Chapter 7) shows, bodies react involuntarily to memory prompts, sounds of aeroplanes, emergency and fire alarms, all of which prompt the 'psychosomatic workings of [the] body' (Muzaini 2015: 110). Stumbling upon memory in the everyday streetscapes in Berlin, Warsaw and Singapore prompted Drozdzewski (Chapter 2) and Muzaini (Chapter 3) to reflect upon the bodies of kin in those places during the Second World War, drawing connections between past and present. Such reflections have the capacity to stimulate an urge to move out of place. In conjuring the placement of the body in the past, De Nardi (Chapter 6) articulates how her participants' bodies challenged normative assumptions of where bodies dwelled during the war in Italy.

In asking how the body responds to memory, our collection thus presents the body as central to memory, clarifying how identities are enacted and performed through the body. Recently, Longhurst and Johnson (2014: 273) asked if the '"the body" had become little more than a ubiquitous marker of identity and difference'? Clearly, within this collection and the field of memory studies more broadly, the body has been pivotal to how scholars challenge the status quo of representation in memory. Further, the use of the body, in particular the researcher's body, has featured as a methodological tool for many chapters in this volume. This shift from discourse to practice follows Whatmore's (2006: 603) call for a practice turn that exposes the affective capacity of memory on the body to the researcher.

Non-human agency and memory-making

In a collection that focuses on memory, place and identity, it would seem axiomatic that places of and in memory are core to our discussion. While places of memory can be both figurative and actual, here we dwell momentarily on the material, foregrounding the role built structures of memory play in the practices of memory-making. We highlight how the authors in this collection move beyond talking about memorials, monuments, museums and material structures of memory as representative of memory narratives, consumed by a passive viewer. Rather, our conversations focus on the agency of these non-human technologies in concomitant processes of memory-making, remembrance and commemoration. Each contribution's focus on places of memory thus falls broadly into one of two categories: those directed at specifically designated sites of national significance; and those where the places of memory are embedded in the everyday.

Hagen (Chapter 14) takes us back to places of memory deemed significant to the Nazi Party, particularly those associated with Adolf Hitler. He discusses how places such as the Brown House and the Bürgerbräukeller beer hall were deliberately chosen as stages for Nazism because such places were seen to radiate certain affective atmospheres (see also McCormack 2008: 43). Hagen notes that members of the Nazi Party were hardly subtle in adorning these places with regalia to advance explicit narratives, but primarily his focus is squarely on the capacity of material structures to perform and affect those visiting them. Similarly, those authors working in national memorial spaces also speak to the affective capacity of memorial exhibits to engender certain commemorative atmospheres. Sumartojo and Stevens (Chapter 11) draw from Böhme (1995) in affirming the contribution of 'staged materialities' in the design of places of memory to 'encourage and enclose the sensory aspect of place'. As 'conditioning environments' (Thrift 2008: 43), national war memorials are written into the significance of nation-building identity narratives. The places, the buildings and the exhibit rooms 'are already rendered with a particular meaning and significance'; they propagate sacredness and reverence, and as Dittmer and Waterton (Chapter 10) go on to argue also engender, 'a power to *move*' and be drawn into a display.

Other authors in this collection have highlighted the affective capacity of memory-markers embedded into, or otherwise part of, everyday places. Framing this discussion of places of memory is the everyday context juxtaposed against 'particular landscapes encoded with . . . preferred readings' (Young and Light, Chapter 4). In such places, events occurred during a period of war and/or conflict; people were killed or hurt; battles may have ensued and atrocities may have taken place. These sites are now marked, remembered and/or adorned with signs of what happened. In contributions by Young and Light (Chapter 4) and Azaryahu (Chapter 5), material reminders of an event in place provide noticeable prompts that bring past into present amid contemporary landscapes. Yet in both chapters, the monuments in question – the 'impaled potato' in Piaţa Universităţii in Bucharest and the relics of Sha'ar HaGai in Israel – have contested histories, the authenticity of which is questioned by varied publics and viewed as stagings of memory. One of Young and Light's participants contends that 'a better monument would not even need to be a monument . . . There are bullet holes in the walls of the National Museum of Art.'

Reminders of past conflict thus are often parts of the same landscapes that, post-conflict, take on quotidian qualities. But because they can be as small as bullet holes they are often missed, unless and until their affective capacities are recognised, known or stumbled upon by passers-by. De Nardi (Chapter 6), for example, chose to consider an inn that had been the 'site of a petty hate crime against civilians', where female guests had their heads shaved by intruding Fascists. Memory of that event is not marked by a memorial, but rather lives on in the narratives and recollections of elderly Italian participants in De Nardi's study. For those involved, the inn serves as a reminder of wartime events and conflicts, even absenting other physical traces. Similarly in Chapter 7, Birdsall conveys how participants recalled that the public archive building in which their interviews with Birdsall took place had been the Gestapo headquarters during the Second World War; again, this exemplifies how events in a place hold through time and emerge in narrative without reference to or need for specific memory-markers. Places and their built structures are, then, only part of an assemblage of memory-making.

Physical traces such as bullet holes or other small-scale memorials are also part of the urban fabric, and over time they fade into that fabric: its bus stops, market halls, in front of or built into residential buildings. In such light, Muzaini (Chapter 3) and Drozdzewski (Chapter 2) both show how encounter is important to thinking through the affective capacity of memorials 'hidden in plain sight' (Tyner et al. 2012: 854). What is striking about these sites as non-human agents in memory-making is that we are more likely to stumble upon them as part of our daily routines than we are to search purposefully for or visit them. The spontaneity of such encounters means that we are rarely primed to be thinking about war and conflict while waiting for a bus or ordering our morning coffee. Thus, unlike visits to a war memorial, absent is what might be called the 'schematic narrative template' (Wertsch 2008; see Dittmer and Waterton, Chapter 10, this volume), and while a given memorial itself may provide clues as to how to be 'read', place-based contexts and everyday surroundings also enthuse memory-making. Both Muzaini and Drozdzewski muse on the unnerving sensibility of place, asking how a place where they have stood could have witnessed such atrocity in the past. These and other chapters in the collection progress scholarship on how the everyday landscapes of post-war cities have, hold and (re)create memory.

Structuring memory, identity and place

The volume is organised around three thematic parts: (Part I) Placing memory in public; (Part II) Narrative memorial practices: storytelling and materiality in placing memory; and (Part III) Commemorative vigilance and rituals of re-membering in place. Each part has been designed in response to the efforts of our contributors to advance theoretical, methodological and empirical debates around memory, identity and place, and to consider, in particular, how these discussions intersect within the context of war and conflict. While each chapter sits comfortably against those gathered together within these respective parts, most also speak in some way to those in other parts. In order to situate our contributions and draw out a series of imperatives that might deepen our understanding of memory, this introductory chapter has sought to outline and introduce the volume's content, and to make a productive contribution to what continues to be a flourishing scholarship. Key to our contextualisation of memory – and a significant point of departure from other collections dedicated to this subject – has been our focus on the geography of memory, or those *encounters between* studies of war memory and geography. Here, the role of place and the 'body as place' have been pivotal heuristics for thinking through memory's integral role within processes of modern nation building and identity formation. To extrapolate this viewpoint, our three parts represent cogitations on these themes as follows.

Part I: Placing memory in public

In Part I, our four contributions explore how places of wartime memory are often overlain by 'the everyday'. Essays by Drozdzewski (Chapter 2), Muzaini (Chapter 3), Young and Light (Chapter 4) and Azaryahu (Chapter 5) examine (re)productions of wartime memory in plaques, monuments and relics. In each chapter, the authors explore how these memories and events resist forgetting by becoming part of everyday landscapes; that is, they are not designated solely as sites of collective commemoration and instead become part of the urban façade – evident on the sides of buildings, embedded in pavements, on the sides of roads and in place names. These memoryscapes remind us that crimes utterly gruesome in nature occurred in mundane places – sites where we now shop, work or walk, ordinary objects and places, as Byrne (2009: 235) points out, that 'can trigger real pain'. As Byrne goes onto argue, 'we know how these objects and places can lie in quiet ambush for us as we move gingerly across the terrain of each new day' (2009: 235). Their design and spatial orientation affect how they are interpreted as sites of memory, and influence whether and how the public interact with them on a daily basis. Such interactions may be associated with mobility through or to these sites of memory, and also occur through the discursive practices of narrative (re)production.

Drozdzewski draws on two memory installations in the post-war cities of Berlin and Warsaw. Her examination of the small-scale memorial projects – Stolpersteine in Berlin and the Tchorka Tablets in Warsaw – focuses on their placement within everyday landscapes. In each location she observes differing levels of encounter with memory, including her own, and considers how the passers-by position the memorials' significance within nation-building narratives. In another post-war city, Muzaini also examines a small-scale memory project embedded in the everyday. The open-book-shaped plaques, established as part of the fiftieth anniversary of the end of the Second World War in Singapore, mark the memory of past events. Muzaini unpacks his personal journey along a self-choreographed memorial trail. His journey adds weight to our contention that non-human elements in the memoryscape have agency as he unveils how these moments of memory lead him to a greater 'sense of closeness to the nation'. By reference to Bucharest, Romania, Young and Light explore the contested geographies of memory of the Romanian revolution. Using landscape analysis, media and archival sources and interviews, Young and Light seek to better understand how the physical structures of memory in public are remembered and conceived of by the public. In doing so, they sift through the intersections of official and public memory-making, paying close attention to personal and individual responses, bodily practices and performances.

Part I's concluding chapter by Azaryahu focuses on memorialisation at sites where an event of war took place: the Israeli War of Independence (1947–1949). The remains at Sha'ar HaGai of vehicles en route to the then-besieged city of Jerusalem link together history, memory and territory, reinforcing the articulation of homeland identity. Azaryahu considers the affective capacity of the relics in the public place of the highway, reasoning that the relics are witness to the events that took place there, but that their transformation and conservation, and their inability to 'speak for themselves' mediates their symbolic capacity as markers of war and identity.

Part II: Narrative memorial practices: storytelling and materiality in placing memory

The contributions in Part II agitate for the acknowledgement of, and engagement with, the materiality of storytelling and its implications for narratives and memorialisations of war and conflict. De Nardi (Chapter 6), Birdsall (Chapter 7), Shimada (Chapter 8) and Johnson (Chapter 9) all show how explorations of the affective geography of warscapes require that we attempt to make sense of these places and the identities interwoven through differing media – story, sound, art and sculpture. Their focus on storytelling and on the materiality of war memory demarcates a shift beyond the semantic into the realm of the more-than-representational and affective. Materiality offers a useful toehold into the multifarious and elusive worlds of conflict experience: war mementoes, art, sculpture and relics may symbolise and transmit grief, loss and mourning, but they do so only if and when they are in conversation with affective practices of remembering events, people and places (see also Saunders 2000).

De Nardi's contribution interrogates storytelling practice among Italian Second World War veteran fighters to locate who was perceived to be the 'enemy' and who fought whom. In De Nardi's chapter, place features as the locale where dwelling practices ensue and senses of place are mediated by the assumption of appropriate behaviours in those places. The memories uncovered by her participants show how the war disrupted not only Italian (national) dwelling narratives and enacted identities, but also permeated the daily practices and places of her storytellers. Drawing on Pierre Janet's tripartite paradigm of memory – habit memory, narrative memory and traumatic memory – Birdsall explicates processes of Othering in the memory of National Socialism. Birdsall uses sound as an aide-memoire, exploring how her participants remember the war synchronously with their own childhoods and their German cityscapes. An intertwining spotlight weaves through identity, alterity, memory and affects to add a further layer to the tapestry of memory presented in this collection. De Nardi's and Birdsall's chapters most directly engage with the materiality and sensoriality of storytelling. These two chapters critically unpack the affective matter of storytelling by focusing on the sense of hearing and its embodied significance for the memory of the war (Birdsall) and on how identity-making draws on local dwelling perspectives to make places and to understand and place the self and others (De Nardi).

Shimada's chapter contextualises Divis Mountain as a constant amid the changing geopolitics of place and identity in Northern Ireland. She explains how local landscapes of the Troubles are narrated, experienced and embedded through Divis in local memories and identities of her participants. Storytelling as place-making practice and landscape imaginings as a form of cultural resistance are central themes of Shimada's chapter, unpacking the bitter aftermath of the Northern Irish guerrilla wars. The Herald of Jericho, a statue of an angel in the former British-controlled correctional facility constructed of the ruined material of that prison, is a potent locus of embodied memory bearing witness to the enduring materiality of local identities. The Herald of Jericho serves as a site of memory and a site of affectual engagement strongly connected to historical events that shaped local people, stories and places.

Finally, Johnson turns our attention to a form of war commemoration that she argues 'has received limited scholarly attention' amid the wider gambit of scholarly work on war memory. During the First World War, art was a principal means of visually communicating; images of the war commissioned by wartime artists became the remembered images of the event. Johnson focuses on the work of two Irish artists, Orpen and Lavery, as she delves into how the places of war depicted in their paintings were mediated by the affective and emotional engagements each artist had with their own (national) identities during a time when Irish and British relations were fraught with tension.

Part III: Commemorative vigilance and rituals of remembering in place

The five contributions collected together in Part III illuminate a focus on places in memory, distinguished from places *of* memory or commemoration. These places in memory may be constituted by physical locations in the world as well as emotions, associations, 'what ifs', events and the bodies of the storytellers themselves. By focusing the analytic lens on practices of commemorative vigilance and rituals of remembering, the authors in this part draw attention to the more-than-human dimensions of places in memory. Thus, in this third and final part of the collection, the focus is very much on how commemoration is enacted and how places in memory facilitate and elicit the body to remember. An overarching theme of this part is understanding memorial practices as assemblages about which Waterton and Dittmer (2014: 123) posit that 'assemblages are conceived as entities composed of heterogeneous elements irreducible to their role within the larger assemblage'. In memorial places, these elements have the capacity to affect how we remember; that is to say that non-human elements in memorials landscapes play a role in the process of remembering.

At the Australian War Memorial, Dittmer and Waterton (Chapter 10) concentrate on the commemoration of the Kokoda campaign in the Second World War. Using auto-ethnographic approaches, they move through this exhibit – this place in memory – sensing the different parts of the exhibit and extrapolating how they mesh with a wider narrative schematic of Anzac (Australian and New Zealand Armed Corps) memory in Australian national identity. Dittmer and Waterton reason that while the narrative template of Anzac habituates a certain response from visitors to the Kokoda exhibit, namely 'humour, good will and mateship', they also argue that the affect of the exhibit itself must be considered independently from its narrative representation. Sumartojo and Stevens

(Chapter 11) also move through a place in memory encoded with national significance. These authors each attend an Anzac Day service – in Melbourne at the Shrine of Remembrance and in Canberra at the Australian War Memorial. Their focus is on how their bodies in place are 'part of the ritual', how they sense, feel and react to the event as it progresses. They sketch out how atmospheres of Anzac are produced, experienced and encountered through a focus on the more-than-human elements of the commemorative landscape – the lighting, the sounds, the lack of sounds, and the architecture. Each non-human part of this assemblage is discussed as energising the experience. The use of light, sound and material design also takes central place in Witcomb's chapter (Chapter 12) on First World War commemorations unfolding in Australia. Again adopting auto-ethnographic approaches, Witcomb narrates her encounters with two exhibit spaces, teasing out how emotions are currently being used to shape our response to the legacy of the First World War. While Witcomb draws attention to how the exhibits are constructed to elicit certain responses from the audience, she is particularly concerned with the political imperative embedded in these memorial constructions.

The political duty of remembrance is also explored by Wilson's analysis (Chapter 13) of witnessing and affect in modern Britain. Wilson explains that with practices of remembrance of the Great War now passing out of living memory, the role of the witness is foregrounded as crucial to the continuation of war remembrance in the present day. Yet witnessing is fraught with obligations and requires the witness to recognise the effect of the war second-hand, often through normative and rehearsed practices and process of commemoration. Hagen (Chapter 14), meanwhile, takes us into the familiar yet unsettling realm of Nazi propaganda architecture and asks how the Nazi Party attempted to 'create' places of commemoration, and force the propagation of Nazi identity through parades, rallies and bodily practices and performances. Hagen explores how elements in the assemblage of Nazi Party propaganda required the body to move as part of a memory practice designed to withstand the test of time.

In writing this introductory chapter, we hope we have done justice to the innovative ways in which the authors in this collection have continued to think outside of the methodological box, incorporating novel and at times surprising methods to think through the capacity for human and non-human agency in memory narratives and places in memory. Our focus on war and conflict remains crucial to our ruminations on memory, place and identity. We know that war and conflict continue to influence nation building and that we are beckoned as individuals to engage with both our bodies and minds in constructed commemorations of war. The chapters in our collection demonstrate the need to also think forward: What are the possibilities for remembering war into the future? Can a greater appreciation of how memory and identity are linked have spatial manifestations that engender different responses by nation states to war and conflict? What we can say for certain is that the task of understanding memory, identity and place together is critically important at the present, and is forecast to remain so – for the public, for intellectual work, for policy and to aid diverse communities of place and interest to reflect on the significance and heritage of conflict.

References

Ahmed, S. 2010. Happy objects. In M. Greg and G.J. Seigworth (eds) *The Affect Theory Reader*. Durham, NC: Duke University Press, 29–51.

Bal, M. 1999. Introduction. In M. Bal, L. Spitzer and J.V. Crewe (eds) *Acts of Memory Cultural Recall in the Present*. Hanover, NH: Dartmouth College, vii–xvii.

Barad, K. 2003. Posthumanist performativity: toward an understanding of how matter comes to matter. *Signs: Journal of Women in Culture and Society* 28 (3): 801–831.

Barad, K. 2007. *Meeting the Universe Halfway*. Durham, NC: Duke University Press.

Bell, M.M. 1997. The ghosts of place. *Theory and Society* 26: 813–836.

Birth, K. 2006. The immanent past: Culture and psyche at the juncture of memory and history. *Ethos* 34: 169–191.

Böhme, G. 1995. Staged materiality. *Daidalos* 56: 36–43.

Brockmeier, J. 2002. Remembering and forgetting: narrative as cultural memory. *Culture & Psychology* 8 (1): 15–43.

Butler, J. 1993. *Bodies that Matter*. London: Routledge.

Butler, J. 1997. *Excitable Speech*. London: Routledge.

Butler, J. 2004. *Undoing Gender*. London: Routledge.

Byrne, D. 2009. A critique of unfeeling heritage. In L. Smith and Akagawa, N. (eds) *Intangible Heritage*. London: Routledge, 229–252.

Cloke, P. 2002. Deliver us from evil? Prospects for living ethically and acting politically in human geography. *Progress in Human Geography* 26: 587–604.

Curtis, K. 1999. *Our Sense of the Real: Aesthetic Experience and Arendtian Politics*. Ithaca, NY: Cornell University Press.

De Nardi, S. 2015. When family and research clash: the role of autobiographical emotion in the production of stories of the Italian civil war, 1943–1945. *Emotion, Space and Society*. Available online: www.sciencedirect.com/science/article/pii/S175545861 5300086, last accessed 15 July 2015.

Diken, B. and Bagge Laustsen, C. 2005. Becoming abject: rape as a weapon of war. *Body and Society* 11 (1): 111–128.

Drozdzewski, D. 2014. Using history in the streetscape to affirm geopolitics of memory. *Political Geography* 42: 66–78.

Drozdzewski, D. 2015. Retrospective reflexivity: the residual and subliminal repercussions of researching war. *Emotion, Space and Society*. Available online: www.sciencedirect.com/science/article/pii/S1755458615000158, last accessed 24 April 2015.

Dwyer, O.J. and Alderman, D.H. 2008. Memorial landscapes: analytical questions and metaphors. *Geoforum* 73 (3): 165–178.

Halbwachs, M. [1926] 1992. *On Collective Memory*. Chicago: University of Chicago Press.

Hill, L. 2013. Archaeologies and geographies of the post-industrial past: landscape, memory and the spectral. *Cultural Geographies* 20: 379–396.

Hoelscher, S. and Alderman, D.H. 2004. Memory and place: geographies of a critical relationship. *Social and Cultural Geography* 5: 347–355.

Johnson, N. 1995. Cast in stone: monuments, geography, and nationalism. *Environment and Planning D: Society and Space* 13: 51–65.

Johnson, N. 2014. Exhibiting maritime histories: Titanic Belfast in the post-conflict city. *Historical Geography* 42: 242–259.

Jones, O. 2011. Geography, memory and non-representational geographies. *Geography Compass* 5: 875–885.

Langer, M. 1989. *Merleau-Ponty's Phenomenology of Perception: A Guide and Commentary*. Basingstoke: Macmillan.

Longhurst, R. 2001. *Bodies: Exploring Fluid Boundaries*. London: Routledge.

Longhurst, R. and Johnston, L. 2014. Bodies, gender, place and culture: 21 years on. *Gender, Place & Culture* 21: 267–278.

McCormack, D. 2008. Thinking-spaces for research-creation. *INFLeXions* 1 (1). Available online: www.inflexions.org/n1_mccormackhtml.html, last accessed 7 January 2016.

Macpherson, H. 2010. Non-representational approaches to body–landscape relations. *Geography Compass* 4: 1–13.

Maddrell, A. 2013. Living with the deceased: absence, presence and absence-presence. *Cultural Geographies* 20: 501–522.

Muzaini, H. 2015. On the matter of forgetting and 'memory returns'. *Transactions of the Institute of British Geographers* 40: 102–112.

Nora, P. 1989. Between memory and history: les lieux de memoire. *Representations* 26: 7–24.

Nordstrom, C. 1997. *A Different Kind of War Story (The Ethnography of Political Violence)*. Philadelphia: University of Pennsylvania Press.

Olson, E. 2015. Geography and ethics I: waiting and urgency. *Progress in Human Geography* 39: 517–526.

Rose, G. 1993. *Feminism & Geography: The Limits of Geographical Knowledge*. Cambridge: Polity Press.

Said, E. 2000. *Reflections on Exile and Other Essays*. Cambridge, MA: Harvard University Press.

Salerno, M. and Zarankin, A. 2014. Discussing the spaces of memory in Buenos Aires: official narratives and the challenges of site management. In A. Gonzalez-Ruibal and G. Moshenska (eds) *Ethics and the Archaeology of Violence*. New York: Springer, 89–112.

Saunders, N. 2000. Bodies of metal, shells of memory. 'Trench Art', and the Great War re-cycled. *Journal of Material Culture* 5 (1): 43–67.

Saunders, N. 2004. (ed.) *Matters of Conflict: Material Culture, Memory and the First World War*. London: Routledge.

Taylor, C.A. 2013. Objects, bodies and space: gendered practices of mattering in the classroom. *Gender and Education* 25 (6): 688–703.

Thrift, N. 2008. *Non-Representational Theory: Space/Politics/Affect*. London: Routledge.

Till, K.E. 2005. *The New Berlin: Memory, Politics, Place*. Minneapolis: University of Minnesota Press.

Till, K.E. 2012. Wounded cities: memory-work and a place-based ethics of care. *Political Geography* 31: 3–14.

Tolia-Kelly, D.P. 2004. Landscape, race and memory: biographical mapping of the routes of British Asian landscape values. *Landscape Research* 29: 277–292.

Tuan, Y.F. 1977. *Space and Place: The Perspective of Experience*. Minneapolis: University of Minneapolis Press.

Tyner, J.A., Alvarez, G.B. and Colucci, A.R. 2012. Memory and the everyday landscape of violence in post-genocide Cambodia, *Social & Cultural Geography* 13: 853–871.

Valentine, G. 2001. *Social Geographies: Space and Society*. Harlow: Pearson Education.

Waterton, E. 2013. Landscape and non-representational theories. In P. Howard, I. Thompson and E. Waterton (eds) *The Routledge Companion to Landscape Studies*. London: Routledge, 66–75.

Waterton, E. and Dittmer, J. 2014. The museum as assemblage: bringing forth affect at the Australian War Memorial. *Museum Management and Curatorship* 29 (2): 122–139.

Wertsch, J. 2008. The narrative organization of collective memory. *Ethos* 36 (1): 120–135.

Whatmore, S. 2006. Materialist returns: practising cultural geography in and for a more-than-human world. *Cultural Geographies* 13: 600–609.

Part I
Placing memory in public

2 Encountering memory in the everyday city

Danielle Drozdzewski

Monuments commemorating battle sites and heroic figures are common beacons of national identity in publicly articulated memory. They link historic events of significance to present-day representations and celebrations of national identity. Indeed, an established corpus of literature testifies to the importance of the remembrance of war memory to a nation's identity narratives (Chang 2005; Gillis 1994; Koonz 1994; Lowenthal 1994). National remembrance often seeks to collectivise national memories into an 'imagined' community (see Anderson 1991) so that the portrayal of the 'national past' in monuments and memorials has become synonymous with the symbolic transmission of national identity. Atkinson and Cosgrove (1998: 32), in their explorations of war memorials in Italy, have contended that throughout the nineteenth century, national and imperial identities were cemented in the urban landscape through the strategic planning and placement of large public monuments in 'key metropolitan locations'. The creation of nationally significant commemorative spaces remains important to the continuance of a nation's wartime narratives because they become spatial focal points where commemorations occur, even though they often do not mark the location of where the event being commemorated actually took place.

Sites of memory marking the locations of death, murder and other wartime atrocities are very much part of the vernacular of cities whose wartime struggles were played out in city streets and urban capitals. In these everyday landscapes, the site and performance of memory are often not premeditated but part of the same place where we now live, go to shop, work or just walk. Often such memorials convey more individualised memories in that they denote an individual or a small group of people. The Stolpersteine in Germany and Tchorka Tablets in Warsaw are used in this chapter as examples of such small-scale memorialisation, positioned in their everyday environments. Their modest form and implantation in the streetscape minimalise the commemorative scale in a landscape otherwise brimming with state-sponsored memory projects by directing the lens on one or just a handful of victims. Both memorials centralise the geography of official commemoration(s) – they locate memory to place.

Memory in the (post-war) city

This chapter situates its discussion of war memory, place and identity in the context of the everyday landscape. Elsewhere I have drawn from Winchester et al. (2003: 35) to define the everyday as relating to the 'ordinary landscapes in our daily routines', places such as streets, shopping centres, parks and public squares (Drozdzewski 2014a). This chapter focuses on how the legacy of war, fought not in fields or trenches but in the everyday spaces and streets of the city, is articulated, (re)produced and remembered in the material fabric of the city, in its pavements, buildings and with the people who inhabit these places. By focusing on the everyday I seek to better understand how memory intersects (or not) in the daily routines of the people in cities, where war has left an indelible mark. In doing so, this chapter also provides a critique of encounters with memory that have been concretised into material form to create sites of memory in public spaces that spatialise and localise memory to the specific places where an event occurred.

While we often think of memory as an individual mnemonic device, something that conjures or triggers thoughts, feelings and emotions in our minds and our body, memory is also readily imprinted in material form in the fabric of a city. The built urban form of the city houses memories; it can act as a witness to the vicissitudes of everyday life – the mundane and the catastrophic. Public places are thus important commemorative atmospheres (see Anderson 2009), where the portrayal of a nation's memories is entrenched in not only the physical materiality but also the ambiance of the city. As Light et al. (2002: 135) have argued, memory manifests in the 'raising of statues, monuments and memorials' to commemorate 'significant events or personalities from the past' in the visible public 'construction of a shared "national" history'. For Hutton (1988: 315), commemoration is 'a mnemonic technique for localising collective memory', which makes a national narrative visible in public space. Such markers in public space provide points of connection, which as Hayden (1995: 46) has suggested 'trigger memories for insiders, who have shared a common past'. Osborne (2001: 51) has argued that public monuments can act as 'consensus builders', which become 'focal points for identifying with a visual condensation of an imagined national chronicle'. As 'focal points' and historical anchors, monuments link articulations of national identity to a collectively imagined past and through to contemporary generations (Cooke 2000; Foote, et al. 2000). Furthermore, Hayden (1995) and Jacobs (1999) have shown that urban places comprise the many historical layers of identity politics, which factor into how places are conceived in the present and how attachments to place by different groups of people are diverse and fluid through time. Much research has explored how memory is represented, and manifests, in urban and/or city landscapes, especially through monuments, street names, museums and memorials (see, for example, Azaryahu and Foote 2008; Dwyer and Alderman 2008; Hay et al. 2004; Young 1989).

This chapter's key point of departure from this existing literature is that it draws on research that explores engagement and encounters, by the everyday populace, with memories of war embedded *in* the everyday urban façade. These analyses of everyday memorialisation of the Second World War use two capital cities – Berlin and Warsaw – whose urban form was largely obliterated by the end of the war, and for whom (re)construction of the city has meant rebuilding with places and sites of memory that memorialise this obliteration, and the reasons for it. Both cities have highly memorial-laden landscapes, with state and other religious and minority groups invested in building memory into place (Drozdzewski 2014b; Pain and Staeheli 2014). Some of the sites of memory were built with nation-scale commemorative vigilance in mind (see also Nora 1989), such as the Tomb of the Unknown Solider in Plac Marszałka Józefa Piłsudskiego, Warsaw, or Neue Wache on Unter den Linden, Berlin; others, such as the Stolpersteine and Tchorka Tablets, were initiated by individuals for the commemoration of the citizens of those cities. Here, memorialisation spatialises remembrance, affixing it to certain national narratives *and* to specific places, while also buttressing it temporally to certain moments in time deemed pivotal to a nation's identity. My intrigue with these smaller sites of memory is fuelled by their individual and small-scale character and how their form and location potentially influences how the public interact with them (or not), on a daily basis. I see these interactions as a proxy for how memory is not only interpreted but also encountered and engaged with, outside of official commemorations or designated memory sites, by different generations of the public with different temporalities to the events being commemorated. In this vein of better understanding both how memory is linked to the place where it is (re)produced and the people who frequent these places, my research also seeks to respond to Young's (1993: 2) contention that: 'By themselves, monuments are of little value, mere stones in the landscape. But as part of a nation's rite or the objects of a people's national pilgrimage, they are invested with national soul and memory.'

This chapter, then, examines encounters and experiences with small-scale memorials in the context of their design, spatial orientation, how they are positioned within the wider gambit of the city's other memorials, and their effect on the everyday populace, and on me. Like Johnson (1994, 1995, 2002: 293), I take the position that public memorials and monuments should not be conceived merely as a 'material backdrop' from which a story is told but as integral parts of the story of remembering and commemoration. Public monuments can become palimpsests of memory that are inscribed with many layers of meaning (Auster 1997). However, reading the palimpsest is rarely straightforward because interpretations may differ or they may wane in significance with increasing temporal distance from the event. Monuments, and our interpretations of them, are not static entities. Indeed, as Hay et al. (2004: 212) have noted, the 'social meaning of monuments and places are not fixed . . . they are open to change and reinterpretations'. By (de)constructing how memory is (re)produced and transmitted, and then encountered through our everyday mobility and tactility with and through public spaces, geographers can expose the intricacies between memory, place and identity. This chapter's analysis of these linkages considers how we affectually engage, encounter and experience public representations of memory, through and with our everyday mobilities in, on, over and at these memory sites.

To explore these encounters, I draw from ethnographic research that used a qualitative mixed methodology to examine engagement and encounter with specific representations of the past in the suburb of Mitte in Berlin, Germany and in the Śródmieście (the suburb of the city centre) in Warsaw, Poland. At the study sites, participant observation, audio- and video-ethnography and vox pop interviews were combined to operationalise memory-methods in place. At each site, a field diary and GoPro HERO 3+ were used to record encounters with the specific memorial; I wanted to notice things about the stones and tablets and the mobilities of people around them that might otherwise seem ordinary and mundane, or pass my direct attention. During the observations I also recorded field notes of my own happenstances in the memory spaces as well as counting and documenting the types of encounters by passers-by. Following a period of observation, vox pop interviews were undertaken at each field site. Five questions were asked to every respondent (85 in Warsaw and 53 in Berlin) pertaining to reasons (or not) for stopping, links to national identity and to the everyday setting. By teaming the questions with observational methodologies, I sought to direct the spotlight of war commemoration away from the large-scale meta-narratives that so commonly define commemorative vigilance; rather, my aim was to illuminate two very small-scale examples of war memory thickly embedded in their respective everyday streetscapes. In so doing, my intention was to consider how the place of war memory, and its significance to wider national narratives, affects how these specific war memorials are cogitated by the everyday public.

The remainder of this chapter is divided into three sections. First, I detail the Berlin study site. I discuss the example of the Stolpersteine and the interactions with memorials that took place there. Second, I examine the Tchorka memorials in Warsaw, similarly detailing this example of memorialisation in the everyday streetscape. In the third and final section, I provide a brief synthesis of these findings by critically reflecting on the influence of the everyday setting and the focus on the individual.

Stumbling on memory in Berlin's streetscape

There is a significant corpus of work on memory-making in Berlin (see, for example, Åhr 2008; Brockmeier 2002; Karen 2001; Stangl 2003; Till 1999). Till's (2012b: 6) articulation of a wounded city as one that has been 'harmed and structured by particular histories of physical destruction, displacement, and individual and social trauma resulting from state-perpetrated violence . . . over a period of many year[s]' encapsulates Berlin, as does Timothy Ingold's (1993: 153) beautifully phrased conceptualisation of memorial landscapes as being 'pregnant with the past'. The memory-work undertaken in Berlin to memorialise this traumatic past has included a number of significant large-scale memory projects including the Monument to the Murdered Jews of Europe, the Topography of Terror, Neue Wache and the Jewish Museum. Juxtaposed amid these state-sponsored representations of the past is the Stolpersteine project, which Cook and van Riemsdijk (2014) have contended differs by attempting to bring a human dimension back to Holocaust memorialisation in Berlin. The *stolpersteine* are small brass plates, 10 × 10 centimetres in dimension, embedded in the pavements of the streets of Berlin (as well as in 17 other European cities)[1] (Figure 2.1). In Berlin, there are approximately 5,000 stones. Each stone is attributed to a different former resident of the city, recalling their fate under Fascism. Each individual's history is fixed to the location to their last known residence in the city – 'ONE PERSON, ONE NAME, ONE FATE' (Demnig 2015). Gunter Demnig, the artist who instigated the project in 1993, and continues to be solely responsible for laying the stones, cites the Talmud when outlining his motivation for this project, saying that 'a person is only forgotten when his or her name is forgotten' (Demnig 2015). Demnig's intention has been to create reminders of individuals whose names, amid Berlin's larger more all-encompassing memory projects, are forgotten, lost as one of the many numbers of those murdered. Thus 'the material [of the Stolpersteine acts as] . . . a stumbling block preventing the past from being forgotten' (Muzaini 2015: 104). The term *stolpersteine* translates in English to stumbling stone; the intention in adopting this nomenclature was to indicate that the stones were very much conceived as something that we are supposed to stumble upon in the urban environment. Deming also notes that the stone's

Figure 2.1 Stolpersteine at Gipsstraße 3, Mitte, Berlin.

Source: Author.

positioning and their size were strategic elements of their design (personal communication 2014). When they are noticed, their placement in the pavement necessitates that the viewer stops and looks down, bowing their head in a familiar stance of reverence.

I undertook fieldwork on three different streets: Gipsstraße 3 and 4; Ackerstraße 1; and Große Hamburger Straße 30 and 31. The fieldwork consisted of a mixed method of participant observation, audio- and video-ethnography and 53 vox pop interviews and 2 unstructured interviews with Stolpersteine staff in Berlin. The streets were chosen based on their business, their residential character, that they were close to parks and local cafes (as important vantage points) *and* that multiple stones were located at these sites (4 at Gipsstraße, 4 at Ackerstraße, and 11 at Große Hamburger Straße). Their everyday arena means that movement through these streets (and others) in Berlin affords an opportunity to interact with the 'how that same environment was encountered in the past' (Degen and Rose 2012: 3279). The stones become part of the local neighbourhood where the event took place; they have the potential to bring the past into the present, to generate an affective and tactile memory in that our encounters with them involve our mobility to them, over them, onto them and/or past them. As Lisa Hill (2013: 391) has argued, 'walking [and I would add mobility in more general terms] creates powerful recollections because it provokes a distinct and familiar tactility with the world'.

Contrasting the larger memory projects in Berlin, which provoke aspatial remembrance, the stones locate their memorialisation in the place where an event has occurred. In my vox pop interviews, I used the example of Monument to the Murdered Jews of Europe (at Potsdamer Platz), as an exemplar of a state-sponsored memory project, because its scale juxtaposes the Stolpersteine in terms of size, target audience and its aspatial mandate, but crucially to this study its scope of representation focuses attention away from individuals and towards collective trauma. The majority of respondents stated that they preferred the Stolpersteine to the larger monument. For example, one stated: 'this [the Stolpersteine] is more personal to me. You are given a name. Whereas if you deal with something at Potsdamer Platz, it's a group of people, and this is individual' (Respondent 104, 3 May 2014), while another responded: 'Stolpersteine are more decent and more personal than the memorial at Potsdamer Platz. I like this more' (Respondent 100, 3 May 2014). These respondents' experiences highlight Jones' (2011: 4) contention that: 'notions of being-in-place are powerful, even fundamental to everyday life . . . Within this, memory is key as it is one means by which people are in place/landscape.' Through their placement in the everyday streetscape of Berlin, the stones afford the opportunity for the passer-by to interact within them and the narratives of the past as part of their daily routines *and* they engender a tactile engagement with memory in that we stumble over, on or upon them. The following sections will explore these two points.

At each set of *stolpersteine*, I watched and counted people's mobilities in, over and on the stones. These mobilities and encounters surprised me. In the planning stages of this fieldwork, sitting at my desk in Sydney, I had had grander expectations of what I was going to see, not in the sense that I was disappointed at the stones themselves but that I had anticipated a greater interaction by the public with these sites of memory. My fieldwork uncovered a different story. At Gipsstraße, over the seven days of observation (consisting of 3.5 hours of observation in 30-minute increments), 255 people passed the four *stolpersteine* on that street. In the same time frame 2 people looked down at the stones and 2 people stopped. At Ackerstraße, 296 people passed, 5 people looked down at the stones, 2 people stepped on them, no one stopped at them. At the busiest site, Große Hamburger Straße, which was frequented by tour groups who stopped spe- cifically at the site, 511 people passed by the stones, 17 people looked down at them, 27 people stepped on them and 21 people stopped at them (this does not include the designated tour groups at this site, which would raise this number to approximately 70 people who stopped).

So what does this tell us? On the surface, it tells us that not many people stop or even look down at the stones. Their compactness in this memory landscape means that more often than not, people did not stop, they walked on by, or over, or on them. Are the *stolpersteine* 'hidden in plain sight' – an adage Tyner et al. (2012) and others have used to refer to landscapes of violent pasts in Cambodia? I reason that there are myriad of other explanations as to why people did not stop, the most cogent being that some people, particularly residents, are already con- scious of their presence in the streetscape and do not look down every time they pass them. This exact point was made by an audience member from Hamburg during a recent conference presentation of this work. The person stated that they knew the exact the locations of the stones in their neighbourhood and while they did not look down when they passed them every day they did consciously think on them as they stepped over them. Yet others may simply not notice the stones, they may be tourists or new to the city or unaware that the *stolpersteine* even exist. As Law (2005: 440) reminds, the 'the street looks and feels differently depending on the perspectives of those inhabiting urban spaces'. Indeed one respondent noted that they thought that the stones were especially for residents, explaining the difference between the Stolpersteine and the Monument to the Murdered Jews of Europe as follows: 'These are two categories. One is more commercial, but good because it represents Germany and Berlin, and show[ed] what happened here. Whereas the Stolpersteine are more for residents and people living here' (Respondent 105, 3 May 2014).

The responses to the vox pop interviews provided further explanations about encounters with these memory spaces. When asked if they stopped and noticed the stones one respondent said: 'if I recognize the name on the stone yes . . . maybe, if the weather is nice, but often one is in a rush' (Respondent 113, 6 May, 2014). Another said: 'only when I had time to stop' (Respondent 107, 3 May, 2014).

As these quotes suggest, the everyday landscape also affords the opportunity for a range of tactile engagements with memory sites, like the Stolpersteine. As I traversed the city during my fieldwork, I often found myself treading on or over the stones; similar to my respondents I did not always notice them, and sometimes I did only after I had initially passed them. What I did record during the observations was that when people did stop, they did dwell even if only fleetingly, they often also looked up at the building behind the stone/s. In this vein, a respondent noted: 'it affects me personally, to stop somewhere is always different than to look at something [on purpose]' (Respondent 95, 2 May 2014). Another commented that on stumbling upon a stone: 'I just read the name and thought how this person would have developed and what they would have done if they were still alive' (Respondent 98, 2 May 2014). This type of commemorative memorial *invites* our engagement with memory due to its placement in places where we may walk, shop and work. The affective capacity of the stones relates also to their tactility; the fact that we touch them or come close to touching them can act to prompt memory. But the stone leaves only the minimum information of each person; the passer-by is left to place the memory, situate it in the context of their experiences, which may or may not include their sentiments towards German nationalism, to totalitarianism, or perhaps their own migrant histories.

I certainly found myself emulating respondents' encounters. At each stone that I noticed, I wondered what this spot might have looked like on the day that that person was removed from their home. Yet, in conjuring these thoughts my feet had a sense of urgency, a sense of wanting not to dwell, but to move on, move away. In these moments, my cognition of what happened in these places was unnerving and upsetting. These feelings meant that even though I was in a very public and open place, my instinct was to move on because the thought of what had happened in that place in the past was unsettling. In those moments I could most poignantly relate to Bergson's assertion of memory that: 'there is no perception which is not full of memories' (Bergson 1911: 24). Retrospectively then, and drawing further from Halbwachs ([1926] 1992), my ability to interpret the memory of the stones in the present day was mitigated and informed by my knowledges of the events that precipitated the deportation of these Berlin residents. Till (2012a: 22), in her response to the dialogue that followed her 'Wounded cities' article in *Political Geography* stated that: 'Trauma does not occur from an event or occurrence that caused pain or suffering per se, but from an individual's inability to give the past some sort of story.' My knowledge of the past, my story, was socially constituted and constructed by the discourses I had learnt of the Second World War in the present, which were also mediated by the situated knowledges afforded to me by my Polish family, and my post-memory/ies of this event (Hirsch 1997). I contend that in the streets of Berlin, the wider story of the Stolpersteine is wrapped up with our ability to be mobile through the environs in which they were embedded.

The *stolpersteine* can be missed in the hurriedness of our everyday mobilities, but when we do notice them they remind us of the small and hidden places of memory missed within our cities, the grievous histories. One respondent noted that the 'symbolic meaning [of the stones] is more significant when you step on it' (Respondent 98, 2 May, 2014); another cautioned: 'I try not to step on them' (Respondent 114, 6 May 2014); while another highlighted that: 'most interesting is that you step on it' (Respondent 94, 2 May 2014). These responses provoke us to consider the affect of the tactility of the stone, to think about how their palpable character acts as a memory prompt. These reminders, encountered under our feet through touch, and visually with our eyes, differ in character from forms of engagement in other memory spaces (in Berlin) because when we visit a museum or a memorial it is often a premediated act, we know we are going to a memory space and we have time to process our thoughts about what is being commemorated in those places. With the *stolpersteine*, unless we are purposefully locating a particular stone or partaking in an installation of a stone, more often than not encounters with them are extemporaneous. One respondent explained this interaction as: 'more of a random recognition, by chance I recognise them' (Respondent 92, 2 May 2014). Other respondents specifically noted the influence of the everyday setting: 'Jews were our neighbours, our parent's neighbours. It's important to revitalize memories as they disappear from everyday life' (Respondent 110, 6 May 2014). Others stated: 'It's everyday life. It's everywhere. The mass of stones makes people aware' (Respondent 114, 6 May 2014) and 'this is more everyday. [It] brings attention to the fact that they were taken' (Respondent 115, 6 May 2014). Shields (2012: 15, original emphasis) has contended that 'disappearance does not offer the body such solace. The mind reels to repopulate place and the body is led to turn *without gesture* around a gap, a space of absence that is both a material and abstract representation.'

In Berlin, the stones are that 'gap', which as the above quotes highlight, prompt rumination on the absence of people that were once neighbours in those same places – as Gunter Demnig has intended them to.

Framing memory in the walls of Warsaw

Warsaw is a city with a long history of foreign occupation and struggle for national autonomy (Drozdzewski 2008). It is a city that has undergone multiple changes in its urban form that simultaneously bear the hallmarks of a city that has been under siege, the relics of a city purposefully obliterated and the lasting foundations of being rebuilt (in brutalist form) by yet another occupying force. Chmielewska (2005: 129) articulates these changes well, noting: 'Warsaw is accustomed to having its memory, and its present, re-framed through the consecutive re-inscribing of its surface by the alternate forces of history: political, linguistic and cultural.' Reminders of this change manifest materially not only in the (re)built form of the city but also in the memory of places in the city. As Orla-Bukowska (2006: 177) has contended: 'in central Warsaw hardly a step can be taken without encountering histories of original edifices on their contemporary replicas, or roll calls of tortured and executed innocents outside prisons'. The city itself – its streets, any existing pre-war buildings, remnants, street names and its residents' memories – is a memorial in its own right. The scale of memorialisation in Warsaw means that one can possibly encounter reminders of the nation's Second World War history at least on every street. The overriding premise of Warsaw's memorials narrates a city victim to the mass destruction of totalitarian occupation, whose memory has been (re)constructed to accord (for the most part) with memorialising well-rehearsed narratives of Polish nationhood. In positioning the representations of Polish memory in this way, I seek not to devalue the content or rationale of the commemorations, rather I aim to draw attention to how they are tightly bound up with and interwoven in distinctly nationalistic discourses. Moreover, I seek to emphasise the story of Warsaw's (and Poland's) history and historical cultural memory, which is very much underwritten in Anglophone literature (Dorrian 2010; Tucker 2011).

Many of Warsaw's monuments are small scale; they can be inscriptions on plaques in buildings and public squares, crosses, on religious shrines or *zabytki* (translated variously as monuments, relics and historical treasures), all of which can also be adorned by candles and flowers. These smaller monuments often commemorate individual events or individual deaths at the hands of both the Nazi and Soviet occupiers of the city – rather than attempting to emblematically represent the suffering of a whole group or a whole event, like the Warsaw Uprising memorial for example. The implication of this level of memorialisation is that such memorials can easily be missed, they can blend into the streetscape and/or become so ubiquitous that they no longer draw the attention of the casual passer-by, in the words of one respondent, 'they often kind of melt with the building and it's hard to notice them' (Respondent 24, May 21 2014).

An example of this small-scale memorialisation in Warsaw is the Tchorka Tablets (*Tablica Tchorkowe*), which were designed by Karol Tchorka to commemorate the loss of Polish lives during the Second World War. There are approximately 200 Tchorka memorials positioned in various locations in the Warsaw streetscape. The memorials began as rudimentary wooden, metal or even cardboard crosses, placed by residents returning to the city in 1945 to mark the sites where either actual bodily remains still lay, or where an atrocity was known to have occurred (Davies 2004). Witness testimonies were used to identify the locations of murder; each location and associated witness testimony was documented with help from the Polish Red Cross (Siwek 2014, personal communication). Adam Siwek (2014, personal communication) from *Rada Ochrony Pamięci Walk i Męczeństwa*[2] (Council for the Protection of Struggle and Martyrdom Sites) noted that 'it was also about documenting the numbers of victims and finding out their personal information, when it was (still) possible'. These more temporary memorials were replaced with small black metal plates with white crosses that provided the details of the events that the memorials were marking (Ciepłowski 1987). Then, in 1948, a national competition was held to create a formalised memorial to be used at each site of memory. Karol Tchorka's design, incorporating the Maltese Cross as a reference to the highest Polish military award for heroism and courage (the *Order Wojenny Virtuti Militari* – the War Order of Virtuti Militari), won this competition.

The incorporation of the cross on the Tchorka memorials provides a context for reading memory in two ways. First, as discerned by the form of the cross, the Polish struggle for freedom is positioned within messianic narratives of Polish suffering. A messianic narrative positions Poland as the perpetual victim in European power struggles and its potential saviour (Walicki 1982).[3] Second, the memorials signpost that people like themselves lost their lives in the struggle for Polish freedom, and that this happened not on a distant battlefield but in the streets of the city. This second point is crucial because of the positioning of the crosses on the walls and remnants of the city's buildings. Localising memory in these places, with the use of the cross, provides omnipresent reminders of the war's catastrophic and intimate influence on the citizen *but* the memorial form implies the importance of conveying this reminder alongside a messianic narrative.

The following sections chart how these Tchorka Tablets are regarded by Varsovians and whether the everyday setting indeed communicates this messianic message. I observed two sites where the Tchorka memorials were embedded in the walls of buildings. Both were busy locations: *Hala Mirowska* (Mirowksi Marketplace) in the Mirów district in central Warsaw; and, at a bus stop at the corner of *ul. Bracka* (Bracka Street) and *Aleje Jerozolimskie* (Jerusalem Avenue). At each site I repeated the method used in Berlin; I commenced the fieldwork with five periods of observation and video-ethnography, followed by five periods of vox pop interviews during which 83 respondents were interviewed.

On each Tchorka Tablet, the common inscription in Polish in the centre of the cross testifies to the sacrifice and suffering of the Poles during the Second World War: 'this place is sanctified by the blood of Poles in their pursuit of freedom for their homeland'. The second inscription positioned under the cross is specific to the circumstance of each plaque's location. It decrees the Nazi perpetrators, the date and the number of Polish victims at that particular place (Figure 2.2).

Figure 2.2 Tchorka Tablet on *ul. Bracka* 16, Śródmieście, Warsaw.

Source: Author.

That both inscriptions are in Polish foremost delineates that the memorial is speaking directly to a Polish audience. Gordon (2001: 16–17) has asserted that 'historical plaques and especially monuments [are used] to guide sentiments of patriotism'. The Tchorka Tablets effectively impart a patriotic sentiment; 83 of the 84 respondents who undertook a vox pop interview affirmed that the memorials were significant to national identity and memory. Notwithstanding the significance of a memorial's potential to communicate messages of national identity, these sentiments of patriotism in a Polish context are very much learnt, (sometimes) inherited and transmitted inter-generationally as well as transnationally (Drozdzewski 2012; Tucker 1998). Ergo, I would have expected the Poles I interviewed to convey this response especially in light of Nora's (1996) suggestion that 'public history becomes internalised by individuals and shared as public memory' (Gordon 2001: 7). Moreover, De Nardi (2014: 3) has reasoned that 'the more we know the past, the less we can experience or feel it'. Perhaps then this possibility of not feeling the past, regularly, is why the interweaving of identity discourses of religious conviction with nationalism is such a common theme in the Polish memorial landscape (Porter 2001). Kubik (1994: 189) has argued that in Poland the cross is used as a 'metaphor of national martyrdom' as a 'sign of defiance towards the Communist regime' and a 'symbol of Poland as the Messiah of nations'. The cross acts as a connecting symbol between past and present, and concurrently it conveys those associated nationalistic discourses of struggle for autonomy against foreign occupation. Graham (2002: 1008) has argued that 'societies create emblematic landscapes – often urban – in which certain artefacts acquire cultural status because they fulfil the need to connect the present to the past in an unbroken trajectory'. This assertion was confirmed by many of the respondents' recollections of the memorials, which was based on their memory of the distinctive shape of the cross on the tablets. For example:

> You can see it's in a shape of a cross. It's somehow adequate for commemorating victims, people fallen defending the country, it suits that purpose and I like it (Respondent 34, 21 May 2014).

> It's in the shape of [a] cross, so it momentarily catches attention . . . some kind of plaque for Home Army or its place of martyrdom. So it catches attention for sure. (Respondent 15, 19 May 2014).

> I generally like it, because there is a cross on it and it has a significant connotation. It [links] with national orders (Respondent 63, 24 May 2014).

Further, it was this symbol that provided a reference point to the narrative of memorial, even when the specific event commemorated at the memorial was not recalled. One respondent's statement is emblematic of this intertwining of messianic and national narratives:

> This cross, the shape of a cross. I remember there are other plaques with such a cross . . . I just don't remember what those are about. I rather remember the

outline (of the plaque) than the inscription. But those are all connected to WWII, [they] commemorate executed people . . .

(Respondent 15, 19 May 2014)

The intent of using the symbolism of the cross to depict national identity is for the cross to act as a referent to individual or national identity (Osborne 2001), and to make a statement about a community or national response to a defining event in a nation's history (Winter 1995). While this message was definitely evident in the responses of the vox pop interviews, results of the observational component shed light on how the Tchorka memorials, enveloped into their everyday surrounds, were being noticed/read (or not) by the people passing them. At *Hala Mirowska*, over the five days (2.5 hours) of observation, 598 people passed the memorial, 10 people looked at the memorial and 2 people stopped at it. At *ul. Bracka*, in the same time period, 890 passed, 10 people looked at the memorial and 2 people stopped at it. To better elucidate these results, which suggest that ordinarily little attention is paid to these memorials, I direct the lens on their spatiality as a way understanding how memory positioned within the urban fabric is encountered and experienced. Like at the Stolpersteine, because the two sites of the Tchorka memorials were in very busy public places, frequented on people's way to shop or travel, it is entirely possible that there is an instinctual 'a-where-ness' (after Osborne 2001) of where these memorials are located, and thus less of a necessity to stop and look at them at every pass.

The majority of respondents indicated that they had stopped at a Tchorka memorial somewhere in the city, *and* the majority were acutely aware of what was being memorialised. Some indicated, for example, 'I've read [it] . . . I walk here every day. I've even lit a candle' (Respondent 3, 19 May 2014) and 'I remember, because I'm walking there often' (Respondent 17, 20 May 2014). The notion of repetition through the space is highlighted here as creating a consciousness of the memorial. In this vein, Hill (2013: 392) has argued that 'repetition is crucial to the reproduction and evocation of memory; the repetition of stories told, objects used, and paths walked'. Hill's suggestion encourages us to think about our movement through places that are familiar to us, and how these places have the power to invoke certain memories. In Warsaw, as previously mentioned, it is possible that all places within the city have the potential to arouse memories of the Second World War; Mayerfeld-Bell (2004: 813, original emphasis) refers to these as the 'ghosts of place' – a *'sense of the presence of those who are not physically there'*. When asked about the everyday setting of the Tchorka memorials, more respondents linked the fact that the memorials locate memory in the place where the event took place than to any other facet of the memorial.[4] The importance of remembering these places of significant sacrifice is highlighted in the following response: 'Warsaw is such a place/city where history has left lasting and huge stigma and I think that Warsaw's residents should remember and know that in these places, where they are walking now, something important happened' (Respondent 12, 19 May 2014).

The importance of localising memory to place in these tablets in Warsaw seemed to have an amplified significance for the Poles I interviewed. Being in the place where you know your kin have been executed has powerful repercussions on the sense of self and of your connection to the nation. I certainly recall and recorded a sense of discomfiture at these sites. In my field notes, I wrote: 'standing in the spots where multiple Poles were killed feels very eerie' (11 May 2104). Returning to the same site one year on, my field notes again recall my uneasiness in place and my attempts to decipher these feelings:

> Perhaps this feeling of being uncomfortable tells me something about the monument that is also conveyed in the fact that they are all mostly [written] in Polish. That is, that you are supposed to know about this history, it is history embedded in place, it is irresponsible not to know it. By stopping I show that I must not know it, that I am foreign, which then begs the question of what would I want to take that picture for anyway. I feel uncomfortable knowing this happened here, I feel uncomfortable that I am interested in the fact of knowing that this happened.
>
> (Field notes, 2 June 2015)

Unlike in Berlin, in Warsaw, the people killed in these locations were of my grandparents' ilk. My field notes also record that 'there is a connection between the general ages of those selling flowers and the fact that they are most likely to be the ones who might remember what the plaque recalls'. The people around me, especially at *Hala Mirowska* selling flowers and used wares, are those with whom I felt most uncomfortable. They are also those for whom the plaque is not merely a representation of what happened in this city's past. Rather, the plaque(s) denote what happened in their past, perhaps to their families, friends, kin – in those places. It was no coincidence then that during the vox pops, it was elderly respondents who regularly dictated the plaque's inscription verbatim, but frequently also provided additional dialogue on the events commemorated there. Often, the recitation of these historical notes invoked an emotional response, which included visible distress, tears, sighing, even though the interviews themselves were by design very brief.

A strong sentiment about the exactness of place was apparent in my own and my respondents' opinions about the memorials' locations. I contend that focusing on the everyday surrounds of the tablets prompts a more nuanced understanding of how memory is encountered in place here. Warsaw is a city undergoing rapid intensification of its urban centre. Amid this changing landscape, these tablets are vestiges of its past. It is highly unlikely that, given the long-standing associations between memories of war and Polish identity within discourses of Polish nationalism, these events would necessarily be forgotten. As the following respondent notes, these tablets have the potential to provide small prompts because they are in places we frequent on a daily basis:

> [The Tablets] are located in places that are visited frequently, on the paths used to commute to work or other errands. Maybe someone [may] get

interested by it and stop to read the inscription. [In contrast] not everyone will go to the Tomb of Unknown Soldier to read what is written there.

(Respondent 17, 20 May 2014)

The material permanency of these tablets means that in Warsaw memories of war tangle the past and the present at bus stops, market places and residences. In the everyday landscape they attempt to defy forgetting, and defy continuity, by underscoring a highly traumatic and difficult past out in the open public spaces.

Encounters and experiences with everyday memorialisation

Memory positioned within the urban streetscape, such as the Stolpersteine and the Tchorka Tablets, differs from state-sponsored memory projects not only in size, but also in terms of locating memory where an event occurred. This localisation of memory means that we encounter these moments of a nation's past in places enmeshed with our daily routines. At these sites of memory we stumble upon memory non-causally, at a bus stop or the entrance to a café. Such stumbling also involves being mobile through, at, over, alongside or on these memory spaces, and it tempers them with different mnemonic qualities. They may unexpectedly spur people to think about the nation's past as they encounter an 'undiscovered' memory site. Equally, our familiarity and habitual movement through the streets of those neighbourhoods may mean that one may step over, or pass something, that they recognise the meaning of – as some of the respondents in this research indicated – but that they do not necessarily stop. Instead, these fleeting encounters act like flickers of memory grounding understandings of the city, the nation and its history.

The Stolpersteine and the Tchorka Tablets provide distinctive frames of reference on national memory. The Stolpersteine respond to a perceived lack of the individual in larger forms of memorialisation in Berlin. The Tchorka Tablets also memorialise a small number of individuals; however, memory is mounted in these tablets together with a well-recognised nationalistic narrative. In returning and responding to Young's contention that related the value of monuments to their propensity to engender a nationalistic response, I assert that perhaps this relationship is more complicated and non-linear than this assumption implies. The *stolpersteine* remind the passer-by of one person's fate amid a dire war; they can impel us to link that exact location where we are standing to the past – that we are connected through our shared geography with the person(s) being removed from their homes. The nation is, of course, present in this dialogue but it is foregrounded by the individual, the person who left the city, and the person standing in the same spot where it happened. In Warsaw, we also share an uncomfortable geography with the past in the locations of the Tchorka Tablets. We are reminded of individual sacrifice alongside the nation's suffering; we are given the tools – the form of the cross and the dialogue on it – to encourage this memory narrative. Yet in each city's post-war landscape, engagement with the past, whether premeditated or spontaneous, need not be facilitated by specific memorials, though they do act as memory jolts. The affective atmospheres of each city's encounters with war, trauma and totalitarianism means that our engagements with the past there are not only prompted by material remnants of memory, but also by atmospheres of memory that hang stealthily onto these cityscapes (Anderson 2009). Such atmospheres make feet want move away from spaces where we know awful things happened; likewise, they make us uncomfortable standing in the location where people have been executed. These feelings, sometimes engendered by the tactility of memory, 'occur before and alongside the formation of subjectivity [about specifically commemorated events], across human and non-human materialities, and in-between subject/object distinctions' (Anderson 2009: 78). They happen in spaces of the city where we least expect our daily routines to be interrupted by thoughts of war and our connections to the past.

In each city's post-war landscape there are stories not only of the wounding of the city or the nation (more generally), but also of wounded individuals, once part of those cities. Each memorial prompts us to recall the war's outcome on individuals, in doing so and by way of its location, it also gives this memory a geography. The everyday location of these memorials means that we potentially have points of connection with these narratives of wounding throughout the city. Even though people may not stop at every stone or every tablet – as was the case in this research – small-scale memorialisation in the cityscape warrants our attention primarily because we think it is, and it is, easily missed. But the spontaneous *and* habitual encounters through our cities perhaps have the greatest potential connection with the city's past, through our ordinary and everyday movements. The potential, then, for recognising how a politics of memory is expressed in these memorials, and how it is received by the public looms large.

Notes

1 The Stolpersteine website sets out a list of guidelines for people wishing to have stones installed both in Germany and elsewhere in Europe. In most cases the local council needs to be consulted about the placement of the stones. Some councils, such as Villingen in Black Forest in Germany, have refused to give permission to lay the stones. Until recently the City of Munich had also banned the laying of *stolpersteine*; the opposition to the Stolpersteine project in Munich emanated from a senior Jewish community member who disagreed with the positioning of the stones in pavement, where the memory of those people could be trodden on, dirtied or passed over.

2 The main function of this Committee is the establishment and maintenance of memorials, monuments and war cemeteries. The Committee is part of the Prime Minister's Office.

3 Walicki (1982: 244) has further explained the historical underpinnings of Polish messianism as being 'of great value for the Poles – it enabled them to explain the national catastrophe of Poland, and made them believe that her [*sic*] sufferings were not in vain, since, like the suffering of Christ, they served as a purifying force for the general redemption and regeneration of mankind [*sic*]'.

4 Some of the other coded responses for this question about the everyday character of the memorial included: the ability of the memorial to catch attention; that they needed to be in busy places; that they were important to the Polish nation; and that all types of memorials and monuments are of equal importance.

References

Åhr, J. 2008. Memory and mourning in Berlin: on Peter Eisenman's Holocaust-Mahnmal. *Modern Judaism* 28: 283–305.

Anderson, B. 1991. *Imagined Communities: Reflections on the Origin and Spread of Nationalism*, New York: Verso.

Anderson, B. 2009. Affective atmospheres. *Emotion, Space and Society* 2: 77–81.

Atkinson, D. and Cosgrove, D. 1998. Urban rhetoric and embodied identities: city, nation, and empire at the Vittorio Emanuele II monument in Rome, 1870–1945. *Annals of the Association of American Geographers* 88: 28–49.

Auster, M. 1997. Monument in a landscape: the question of 'meaning'. *Australian Geographer* 28: 219–227.

Azaryahu, M. and Foote, K. 2008. Historical space as narrative medium: on the configuration of spatial narratives of time at historical sites. *GeoJournal* 73: 179–194.

Bergson, H. 1911. *Matter and Memory*. London: Allen and Unwin.

Brockmeier, J. 2002. Introduction: searching for cultural memory. *Culture and Psychology* 8: 5–14.

Chang, T.C. 2005. Place, memory and identity: imagining 'New Asia'. *Asia Pacific Viewpoint* 46: 247–253.

Chmielewska, E. 2005. Logos or the resonance of branding: a close reading of the iconosphere of Warsaw. *Space and Culture* 8: 349–380.

Ciepłowski, S. 1987. *Napisy Pamiątkowe w Warszawie XVIII–XX w.* Warszawa: Państwowe Wydawnictwo Naukowe.

Cook, M. and van Riemsdijk. M. 2014. Agents of memorialization: Gunter Demnig's Stolpersteine and the individual (re-)creation of a Holocaust landscape in Berlin. *Journal of Historical Geography* 43: 138–147.

Cooke, S. 2000. Negotiating memory and identity: the Hyde Park Holocaust memorial, London. *Journal of Historical Geography* 26: 449–465.

Davies, N. 2004. *Rising '44: The Battle for Warsaw*. London: Pan Macmillan.

Degen, M.M. and Rose, G. 2012. The sensory experiencing of urban design: the role of walking and perceptual memory. *Urban Studies* 49: 3271–3287.

Demnig, G. 2015. Stolpersteine. Available online: www.stolpersteine.eu/en/home/, last accessed 23 May 2015.

De Nardi, S. 2014. An emboided approach to Second World War storytelling mementoes: probing beyond the archival into the corporeality of memories of the resistance. *Journal of Material Culture* 19: 443–464.

Dorrian, M. 2010. Warsaw: tracking the city. *Journal of Architecture* 15: 1–5.

Drozdzewski, D. 2008. Remembering Polishness: articulating and maintaining identity through turbulent times. Unpublished PhD thesis in Human Geography, University of New South Wales, Sydney.

Drozdzewski, D. 2012. Knowing (or not) about Katyń: the silencing and surfacing of public memory. *Space and Polity* 16: 303–319.

Drozdzewski, D. 2014a. When the everyday and the sacred collide: positioning Płaszów in the Kraków landscape. *Landscape Research* 39: 255–266.

Drozdzewski, D. 2014b. Using history in the streetscape to affirm geopolitics of memory. *Political Geography* 42: 66–78.

Dwyer, O. and Alderman, D. 2008. Memorial landscapes: analytic questions and metaphors. *GeoJournal* 73: 165.

Foote, K.E., Toth, A. and Arvay, A. 2000. Hungary after 1989: inscribing a new past on place. *Geographical Review* 90: 301.

Gillis, J.R. 1994. Memory and identity: the history of a relationship. In J.R. Gillis (ed.) *Commemorations the Politics of National Identity.* Princeton, NJ: Princeton University Press, 3–24.

Gordon, A. 2001. *Making Public Pasts: The Contested Terrain of Montreal's Public Memories, 1891–1930.* Montreal: McGill-Queen's University Press.

Graham, B. 2002. Heritage as knowledge: capital or culture? *Urban Studies* 39: 1003–1017.

Halbwachs, M. [1926] 1992. *On Collective Memory.* Chicago: University of Chicago Press.

Hay, I., Hughes, A. and Tutton, M. 2004. Monuments, memory and marginalisation in Adelaide's Prince Henry Gardens. *Geografiska Annaler* 86B: 201–216.

Hayden, D. 1995. *The Power of Place: Urban Landscapes as Public History.* Cambridge, MA: MIT Press.

Hill, L. 2013. Archaeologies and geographies of the post-industrial past: landscape, memory and the spectral. *Cultural Geographies* 20: 379–396.

Hirsch, M. 1997. *Family Frames: Photography, Narrative, and Postmemory.* Cambridge, MA: Harvard University Press.

Hutton, P.H. 1988. Collective memory and collective mentalities: the Halbwachs–Aries connection. *Historical Reflections* 15: 311–322.

Ingold, T. 1993. The temporality of the landscape. *World Archaeology* 25: 152–174.

Jacobs, J.M. 1999. Cultures of the past and urban transformation: the Spitalfields market redevelopment in East London. In K.J. Anderson and F. Gale (eds) *Cultural Geographies.* Melbourne: Addison Wesley Longman Australia, 241–264.

Johnson, N. 1994. Sculpting heroic histories: celebrating the centenary of the 1798 rebellion in Ireland. *Transactions of the Institute of British Geographers* 19: 78–93.

Johnson, N. 1995. Cast in stone: monuments, geography, and nationalism. *Environment and Planning D: Society and Space* 13: 51–65.

Johnson, N. 2002. Mapping monuments: the shaping of public space and cultural identities. *Visual Communication* 1: 293–298.

Jones, O. 2011. Geography, memory and non-representational geographies. *Geography Compass* 5: 875–885.

Karen, E.T. 2001. Returning home and to the field. *Geographical Review* 91: 46–56.

Koonz, C. 1994. Between memory and oblivion: concentration camps in German memory. In J.R. Gillis (ed.) *Commemorations the Politics of National Identity.* Princeton, NJ: Princeton University Press, 258–280.

Kubik, J. 1994. *The Power of Symbols against the Symbols of Power: The Rise of Solidarity and the Fall of State Socialism in Poland.* University Park: Pennsylvania State University Press.

Law, L. 2005. Sensing the city: urban experiences. In P. Cloke, P. Crang and M. Goodwin (eds) *Introducing Human Geographies.* London: Arnold, 439–450.

Light, D., Nicolae, I. and Suditu, B. 2002. Toponymy and the communist city: street names in Bucharest, 1948–1965. *GeoJournal* 56: 135–144.

Lowenthal, D. 1994. Identity, heritage, and history. In J.R. Gillis (ed.) *Commemorations the Politics of National Identity.* Princeton, NJ: Princeton University Press, 41–57.

Mayerfeld Bell, M. 2004. *An Invitation to Environmental Sociology.* Thousand Oaks, CA: Pine Forge Press.

Muzaini, H. 2015. On the matter of forgetting and 'memory returns'. *Transactions of the Institute of British Geographers* 40: 102–112.

Nora, P. 1989. Between memory and history: les lieux de memoire. *Representations* 26: 7–24.

Nora, P. 1996. *Realms of Memory: Rethinking the French Past. Vol. 1, Conflicts and Divisions.* New York: Columbia University Press.

Orla-Bukowska, A. 2006. New threads on an old loom: national memory and social identity in postwar and post-Communist Poland. In R.N. Lebow, W. Kansteiner and C. Fogu (eds) *The Politics of Memory in Postwar Europe.* Durham, NC: Duke University Press, 177–209.

Osborne, B.S. 2001. Landscapes, memory, monuments, and commemoration: putting identity in its place. *Canadian Ethnic Studies* 33: 39–77.

Pain, R. and Staeheli, L. 2014. Introduction: intimacy-geopolitics and violence. *Area* 46: 344–347.

Porter, B. 2001. The Catholic nation: religion, identity, and the narratives of Polish history. *The Slavic and East European Journal* 45: 289–299.

Shields, R. 2012. Urban trauma: comment on Karen Till's 'Wounded Cities'. *Political Geography* 31: 15–16.

Stangl, P. 2003. The soviet war memorial in Treptow, Berlin. *Geographical Review* 93: 213–236.

Till, K.E. 1999. Staging the past: landscape designs, cultural identity and *Erinnerungspolitik* at Berlin's *Neue Wache*. *Ecumene* 6: 251–283.

Till, K.E. 2012a. Reply: trauma, citizenship and ethnographic responsibility. *Political Geography* 31: 22–23.

Till, K.E. 2012b. Wounded cities: memory-work and a place-based ethics of care. *Political Geography* 31: 3–14.

Tucker, E.L. 1998. Renaming capital street: competing visions of the past in post-communist Warsaw. *City & Society* 10: 223–244.

Tucker, E.L. 2011. *Remembering Occupied Warsaw: Polish Narratives of World War II.* DeKalb: Northern Illinois University Press.

Tyner, J.A., Alvarez, G.B. and Colucci, A.R. 2012. Memory and the everyday landscape of violence in post-genocide Cambodia. *Social & Cultural Geography* 13: 853–871.

Walicki, A. 1982. *Philosophy and Romantic Nationalism: The Case of Poland.* Oxford: Clarendon Press.

Winchester, H.P.M., Kong, L. and Dunn, K.M. 2003. *Landscapes: Ways of Imagining the World.* Harlow: Pearson/Prentice Hall.

Winter, J. 1995. *Sites of Memory, Sites of Mourning: The Great War in European Cultural History.* Cambridge: Cambridge University Press.

Young, J.E. 1989. The biography of a memorial icon: Nathan Rapoport's Warsaw ghetto monument. *Representations* 69–106.

Young, J.E. 1993. *Texture of Memory Holocaust Memorials and Meaning.* New Haven: Yale Univeristy Press.

3 Personal reflections on formal Second World War memories/ memorials in everyday spaces in Singapore

Hamzah Muzaini

Introduction

Scattered around the island of Singapore, and complementing the myriad bounded spaces of museums, preserved forts and war cemeteries established over the years, is yet another way the nation commemorates the country's involvement in the Second World War and the Japanese Occupation of 1941–1945: the comparably unremarkable markers of history that have been formally inserted into more quotidian surroundings. Among these are the 'open-book'-shaped plaques instituted as part of the fiftieth anniversary of the end of the war in 1995 (Wong 2001), some so blended into the landscape that they can be missed. That is, unless, armed with a proper guide (e.g. NHB 2013), one goes on self-choreographed trails or accidentally stumbles upon them. Although a few are accompanied by material traces (and many of these have already been readapted for other uses), others are nothing more than reminders of what was, memory of past events still seen as noteworthy.

It is with these everyday (although still official) memoryscapes that the chapter is concerned. Unlike the commonly studied museums, preserved battlefields and war cemeteries that are sited within specifically dedicated spaces, everyday memoryscapes are situated as part of public parks, beside bus stops, by roads, on selected buildings, within residential estates, and wherever else one may not usually expect to find them. The chapter first briefly outlines the background of these memoryscapes and how they are used as tools to evoke a version of the war to the public literally at one's doorstep. Drawing upon excerpts from my own recce through these markers, the chapter then examines how they operate as potent triggers of encounters, experiences and memories not necessarily related to the war as represented but just as salient for how one may feel an attachment to the nation. More broadly, it argues for deeper understandings of the complex ways the individual can experience these mundane but still formalised spaces of memory-making.

As commemorative forms, markers such as plaques and storyboards located in everyday spaces may be conceptualised as a cross between the traditional museum and street names that recall elements drawn from the past. Like the former, they provide narratives of history to the public in 'bite-sized' textual nuggets to make more apparent elements in the landscape that are, for the most part, hidden from view if not from knowledge. Aside from narratives, visual aids (such as maps and pictures) are often included to assist readers to imagine the past by juxtaposing the modern landscape with evocations of the same landscape as before, rendering complex geographical and temporal layering of the here and now (Crang and Travlou 2001). As these memory technologies offer the perspective of historical depth, the place is transformed therefore into a palimpsest. However, unlike at conventional museums, where one usually has to escape one's normal routine to spend time learning about history usually behind enclosed spaces, everyday memoryscapes are folded into 'the soap operas of everyday lives', publicly accessible although open to the ravages of the weather (Hebbert 2005: 592). They resemble commemorative street names where '[t]he historical referents coexist simultaneously in the cityscape, but with no linear thread of chronological order connecting them', their salience confined to the information given (Azaryahu and Foote 2008: 183). While remotely positioned memoryscapes may be connected, the information given at each site is relevant mainly to what is available, although, together, they can give a wider bird's-eye view of the particular history of places.

In focusing on these more everyday memoryscapes, the chapter takes emphasis away from the more frequently studied bounded and spectacular spaces of memory as vehicles for collective identity-building and 'think more fluidly about the ways that social memories are constituted throughout society *at different scales and in mundane, everyday places*' (Atkinson 2007: 521, my emphasis). It thus joins the emerging literature that considers less orthodox sites of memory-making, which have often escaped scholarly attention (although see Tolia-Kelly 2004; Hebbert 2005; Butler 2006; Till 2005; Muzaini 2012; Ripmeester 2010). Yet, as this chapter ultimately shows, such memoryscapes, even in their mundanity, too may portend a useful way to 'shift gaze[s away] from a social memory captured within officially sanctioned . . . sites' (Atkinson 2007: 523). While they can at times be so unremarkable as to be neglected, and this too is clearly indicated below, they can also shed important insights contributing to a better grasp of how one *experiences* memoryscapes.

In this regard, the chapter examines how the individual may relate to memoryscapes, or formalised processes of collective identity-building, in less cognitive but more embodied, sensual and affective ways (Clarke and Waterton 2015; Muzaini 2015; Crang and Tolia-Kelly 2010). Crane (1997) writes of the culture of preservation prevalent in contemporary society that removes all vestiges of personal memories in the production of collective memory as a way to retain the integrity of history as 'what has happened' rather than 'what we think happened'. Here, the individual is often written out of the writing/interpretation of history, even as they privilege particular versions of the past to bind a citizenry together and obscure other versions such that while '[c]ertain memories live on; the rest are winnowed out, repressed, or simply discarded [as if] by a process of natural selection' (Yerushalmi 1989: 95). Yet, as scholars have shown, these attempts to create identities by remembrance/forgetting do not always work as individuals do at times renege against moves to appropriate one history by sweeping other histories under the proverbial carpet (Dwyer and Alderman 2008). Regardless of the collective purposes for which such memoryscapes may be put into service, nation-building being one of them, at the end, it is the individual who interprets and accepts or rejects the narratives relayed through them.

However, defining individual experiences simply as the extent to which one embraces or refutes formalised representations of the past as depicted through memoryscapes tends to overlook other ways in which memories (and their attendant spaces) may be engaged with, particularly in terms of emergent encounters, feelings and stories that can (less predictably) occur, especially those that may not be directly tied to the histories as depicted on markers. Indeed, how formal memoryscapes are experienced by each individual is often made complex and contingent upon one's own subjectivities and dispositions at that moment, the materiality of the landscapes and environs to which one comes in contact with, and the temporality during which the experience takes place (Wylie 2005). Thus, despite the task of employing memoryscapes as a way to make members of a citizenry perceive the past in a certain way, *what actually happens when one implicitly and explicitly engages with these memoryscapes may differ not only between individuals but also with the same individual experiencing the same memoryscapes at different times*. While some of these experiences may serve to disrupt intended formal version(s) of the past, others may work to support the purposes for which they have been put into service, such as nation-building. It is with the latter engagements, some fleeting, others more sustained, that this chapter is concerned.

Given the emphasis of this chapter to write the individual back into processes of collective memory-making, it is only logical that, in addition to interviews with key managers of the everyday memoryscapes in Singapore as a means of building a genealogy of how these came to be, autoethnography was the method of choice for the bulk of the chapter. The approach here may be described as 'the process by which the researcher chooses to make explicit use of [his or her] own positionality, involvements, and experiences as an integral part of ethnographic research' (Cloke et al. 1999: 333). A Malay male in my forties, I have no direct relationship with the war, although I have come to be very familiar with, and critical of, how it is commemorated in Singapore (i.e. as tourist guide, as curator of a war museum, as an academic). While I do not claim to be representative of all Singaporeans, as a way to shed light on how engagements with memoryscapes can be highly subjective, as well as to understand the ways in which landscapes, objects and bodies dynamically interact and relate to one another towards particular effects (Waterton and Dittmer 2014), this approach serves my purpose. Indeed, far from any intent to produce the generalisability often associated with quantitative methods, the aim of this chapter is merely to 'open up the tracings of the spatial remains that make us' and 'help trace out the legacies of the past we carry through memory as we practice the present and enter the future' (Jones 2011: 876). Having said that, I have also included, where necessary, voices of other Singaporeans I encountered during the course of a day.

In what follows, the chapter first offers a brief introduction to the everyday markers in Singapore, particularly the ones instituted in 1995 as part of the fiftieth anniversary of the end of the war. It then narrows in on observations (defined as 'moments') made where engagements with three of these everyday memory-scapes, chosen from a total of eight visited on 17 July 2013 for their represent-ativeness of the types of unexpected feelings, encounters and sensations that have made starker my own perceptions about the war and also sense of national attachment. These moments were mined from extensive notes that were taken on the day of my reflections when at these memoryscapes. In doing so, the chapter endeavours to 'relocat[e] the collective back in the individual who articulates it – the individual who disappeared in the occlusion of personal historical con-sciousness' (Crane 1997: 1375), thus refraining 'from a view of the world based on contemplative models of thought and action toward theories of practice that amplify the potential flow of events' (Thrift 2000: 556). After all, as Halbwachs (1980: 51) astutely reminds us, even as a 'man must often appeal to others' remembrances to evoke his own past . . ., one remembers only what he himself [in this case I myself] has seen, done, felt and thought . . . our own memory is never confused with anyone else's'. Finally, this chapter concludes with thoughts on how understanding the affective (as well as cognitive) operations of everyday memoryscapes can contribute towards forging better links between the individual and the place that he/she calls 'home'.

Everyday memoryscapes in Singapore

> Not every Singaporean go[es] to museums. But by putting [the plaques] at ordinary places, we bring history to the people, where they live, where they work . . . Some people think that you don't bother to mark if the building is gone . . . but to us it doesn't matter. As long as there has been significance within that period, it did exercise its own influence and made its contribution, we will go ahead and mark it.
>
> (Sarin Abdullah, Historic Sites Unit, NHB,
> personal communications, 2002)

The Second World War was a significant event in Singapore, then part of British colonial Malaya (which included present-day Malaysia). After about a hundred days of air raids and heavy fighting between the Japanese invaders and the Allied forces, made up of a largely foreign cast,[1] Singapore was plunged, on 15 February 1942, into a dark Occupation period, which lasted for three and a half years until the end of the war in 1945 (Murfett et al. 1999). For reasons already elaborated elsewhere (see Muzaini and Yeoh forthcoming), despite the salience of the event in paving the way for Singapore to gain its independence from the British on 9 August 1965, it was only in the 1990s that it became a part of national historiography,[2] as the proverbial glue to bind the multiracial society as one.[3] Out of the event, three main 'national' discourses were elicited: (1) that Singapore should never have to rely for its protection and preservation on others (emerging out of the sense of betrayal felt against the British for letting the country fall during the war); (2) that Singapore has a history that can be shared among its multiracial population; and (3) that this history is full of tales of local heroism that may be used to inspire younger generations.

Since then, the state has been highly proactive in materialising the war through its landscapes in various ways, where the above narratives are constantly rehearsed: through museums, monuments, preserved battleforts and cemeteries dedicated to different episodes of the war. As part of National Education, Learning Journeys tours too have been introduced where students visit these national memoryscapes as a way to familiarise themselves with the nation's history (and presumably why, as Singaporeans, they need to know it). While these more bounded sites of war memory have become pertinent evidence of the shift in the state's attitude towards the war, from disavowal to embrace, lesser known is another type of memoryscapes emplaced within more everyday locations. While these would include the storyboards put up by the Singapore Tourist Promotion Board in the 1990s, the red plaques of the Preservation of Monuments Board (PMB) to mark their official 'protected' status, or the information boards launched by the National Parks Board at its nature parks to make these places more than just where visitors can enjoy the flora and fauna (see Figure 3.1),[4] the focus of the chapter are the plaques maintained by the National Heritage Board (NHB), especially those introduced in the year 1995.[5]

Figure 3.1 Tourists reading the storyboards at Labrador Park.

Source: Author.

As mentioned, accompanying the fiftieth anniversary of the end of the war in 1995, plaques were installed at 14 locations, selected by the steering committee formed in 1992 by what later became the NHB (established 1993) around Singapore to mark where salient events of the war years took place. At each location, a bronze 'open-book'-shaped memorial was put in place (see figures below). On each are briefly inscribed the specific stories of the sites, alongside a map of 'The Battle for Singapore' and list of all the marked sites (originally 14, although this was recently revised to 20 to include markers emplaced since 1995). The latter information is replicated on all the plaques, thus tying otherwise disparate memoryscapes into parts of a whole, formally 'registering those personalities and events, mythic or real, which have imprinted themselves on popular consciousness' (Samuel 1994: 354; see also Hebbert 2005). The aim of these memoryscapes is to bring memories of the war to the minds of locals and foreigners (*The Straits Times* 9 June 1995), especially Singaporeans not so inclined to visit more bounded sites (as shown by the quote at start of this section).

Since then, the initiative of the 'open-book'-shaped plaques has been subsumed by a larger project, managed by the Historic Sites Unit (HSU), an arm of the NHB. Since 2013, the HSU has been merged with the PMB to become the Preservation of Sites and Monuments Division (PSM). While the red plaques of what was formerly PMB have been left intact as a way to indicate their legal protected status, the others maintained by the former HSU retain the bronze colour as reminders of the (war) past, even as the material sites themselves may no longer be there. While the NHB initiative in 1995 was meant only for war-related sites, PSM's task now extends also to other sites that have played salient roles in the social, cultural, economic and political past of the nation, although non-war related memorials are coloured blue.[6] Given their ubiquity around the country (see Figure 3.2 for locations of the original 14 plaques), it is highly likely that Singaporeans would have encountered these everyday memoryscapes even if they do not directly engage with them.

The reasons behind the decision to have these memoryscapes installed around Singapore were manifold. First, as bits of historical information, they do not take up too much time to read unlike in museums, where more time and effort is needed to visit. Second, where possible, they are located at original locations (with material relics left intact although perhaps readapted to more modern uses) within everyday locations, which allow readers to engage on their own accord with 'the one-dimensional, temporal sequence of historical narrative [although] "draped" across the spatial dimensions of an actual historical site' (Azaryahu and Foote 2008: 180). Furthermore, placing them where war events actually occurred, at or near actually standing material structures that have survived from the war, echoes the tendency of memoryscapes to be located *where a salient historical event took place* to allow visitors to access the 'aura' and sense of standing 'where it all happened'. Third, and especially in cases of the NHB bronze plaques located in residential areas, such everyday memoryscapes can serve to deepen one's awareness (in time and space) of where one is living. Through engagements with these memoryscapes, citizens can be made to become more aware of the nation's history *all around them* in the everyday. Landscapes where they live, work and play

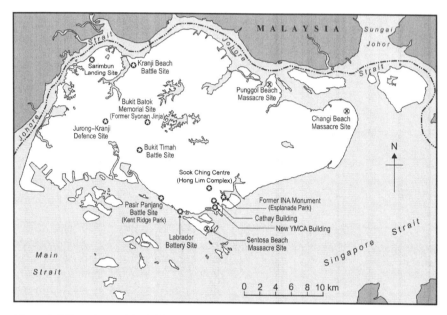

Figure 3.2 Locations of the 14 bronze plaques installed in 1995.

Source: Author.

therefore gain an added dimension – due to the layered histories accrued within them. Fourth, the knowledge gained from reading them also allows for imaginings of the past juxtaposed over what the sites have become. Elsewhere I have written critically of the extent to which Singaporeans have embraced these memoryscapes as representative of the war past of their nation (see Muzaini and Yeoh forthcoming). Here, the focus is on how an individual's engagements with such memorycapes can extend beyond an appreciation/critique of how the past is depicted to include experiences not necessarily tied to the war past but still have the capacity to influence one's sense of self and belonging. To this end, the rest of the chapter reflects some of the emotions, feelings and encounters, aspects that emerged on a day's walk through a selection of these memoryscapes, in particular how these have affected me as an individual and a Singaporean.

Moment 1: Hong Lim Complex

> Today [Hong Lim Complex] is nothing more than a residential area. . . . But by reading the text of the plaque, upon looking up, I could imagine the men being checked by the Japanese for signs of being anti-Japanese . . . the fear they must have felt, even if they are not guilty, given how arbitrary the process was. If Japanese did not like your face, you are dead. It makes me shiver.
>
> (Fieldnotes, 17 July 2013)

Figure 3.3 The 'open-book'-shaped plaque at Hong Lim Complex.

Source: Author.

This excerpt was written about the Hong Lim Complex, a residential/retail building in the heart of what is today the bustling district of Chinatown (Figure 3.3). During the war, the area was, according to the text on the 'open-book'-shaped plaque located there, one of the registration centres of the Japanese Military Police (or the *Kempeitai*). On 18 February 1942, the *Kempeitai* launched a month-long campaign to screen Chinese men between the ages of 18 and 50, and sometimes women and children, for anti-Japanese elements. If they were found guilty, they would be delivered by trucks to outlying parts of Singapore to be executed. It is estimated that tens of thousands of individuals became victims of this exercise known as *Sook Ching* (or 'to purge'). It was certainly one of the most salient war events in the lives of ordinary Singaporeans, which the state felt Singaporeans should never forget, thus meriting the plaque as reminder to Singaporeans not to take peace for granted. According to the then Member of Parliament, Ker Sin Tze (cited in *The Straits Times* 24 July 1995):

> Having a commemorative plaque at this site will evoke painful memories for relatives of victims of the *Sook Ching* operations. But we need to have such a reminder of this dark chapter of our history, so we will never ever forget the price we paid for war.

Two things struck me about my time spent at this particular everyday memory-scape. First, as indicated in my fieldnotes entry above, I was able to, at that moment, layer the contemporary space-time with that of the war years, thus colouring the quotidian present with a more exciting (albeit troubling) past, and 'mak[ing] the present time as deep/complex rather than flat/pure' (Jones 2011: 876; see also Crang and Travlou 2001). Triggered as it was by reading the text on the bronze memorial (albeit too short and dry for my own liking), but complemented as well by my own knowledge of *Sook Ching*, and pictures of the event I have seen, I was somewhat transported back in time, *enchanted* by imaginations of how the 'mundanity of [the] materially ordered space' today (Holloway and Kneale 2008: 303) (sights of men and women patronising the shops and of tourists roaming around Chinatown) was suddenly (and fleetingly) transformed into the past ('of men being checked . . . for signs of being anti-Japanese', an image so strong it made me 'shiver'). Perhaps the scorching heat from the sun that day also contributed to making me feel what it must have been like to stand amidst those who were there on 18 February 1942. Yet, it is key to note that what I imagined and experienced here is perhaps particular only to me based on years of research; someone with less knowledge would probably be affected differently, thus indicating how such memoryscapes are 'constantly in a state of becoming, never truly stable or totally coherent' (Waterton and Dittmer 2014: 124).

Second, I remembered feeling a certain pride from my encounter with the 'open-book'-shaped plaque at Hong Lim Complex. Unlike the everyday memory-scapes that are located in more quotidian locations or in outlying parts of Singapore (see Moments 2 and 3), this particular plaque is in a heritage enclave that tourists visit. In fact, during my time there, I did see a few tourists stand before the plaque taking in the inscriptions about the war. This made me proud that Singapore does have a (hi)story to tell, despite its short existence, beyond the glitz and colours of what is Chinatown today. It also made me proud that Singapore has a government that has, albeit belatedly, deemed it as important to mark such (local) history today as a way to ensure that it is not totally erased, even as there is otherwise nothing there today that brings to mind its past legacy. My engagement with the Hong Lim Complex plaque additionally made me reflect on the peace and tranquillity that Singapore represents today. I do not normally see myself as patriotic in any sense of the word, and I have learnt to always accept the motives and contents of such plaques with a pinch of salt (see Muzaini and Yeoh forthcoming), but on that day at the Hong Lim Complex, I can honestly say I felt content Singapore is a country that looks backwards as well as forwards.

Moment 2: Jurong–Kranji Defence Line

> I am standing at the Jurong West Neighbourhood Park . . . An old Chinese auntie came and asked me what I was doing. I told her about the plaque located there and we ended up having a conversation about her parents' war experiences. She said: 'wah good you tell me . . . I have lived here many years but never knew what that was'.
>
> (Fieldnotes, 17 July 2013)

Another 'open-book'-shaped plaque I visited marks where an arbitrary defence line was established by the British from Kranji River to Jurong River to stop the Japanese advance towards the city from the west (Figure 3.4). According to the text that has been embossed on it, due to miscommunication, the order to defend this line was abandoned by the 44th Indian Infantry Brigade, the 12th Indian Infantry Brigade and the 22nd Australian Brigade that proceeded to retreat further south, thus allowing the Japanese forces to sweep easily through to the city. What was then a swampy area is today a modern public Housing Development Board (HDB) estate, often lauded as 'a success story' where public housing is concerned, which has allowed for a majority of Singaporeans to own their homes (Kong and Yeoh 2003).[7] This mundanity of the residential landscape, where locals can be observed going about their daily routines (e.g. kids playing at the nearby playground, Malay 'aunties' making their way to the market, old men doing *taiji* just beside the plaque) triggered in me a sense of pride. First, it made me proud of how Jurong West (the name of the estate) and Singapore more generally has developed from a rural and swampy setting, populated mainly by Malays, to the modern cosmopolitan city it is today, and by how much more enhanced living conditions here have become (from wooden villages to award-winning blocks of HDB flats). Second, I was proud of the Singapore government for putting the plaque here even though it is an insignificant housing landscape where there is otherwise no trace

Figure 3.4 The Jurong–Kranji Defence Line plaque.

Source: Author.

of what happened before. While the plaque is included in NHB's latest war trail (NHB 2013), no tourists ever make their way here and, from what I observed being there, local residents do not care much for it. The plaque seemed to effectively blend into the background, although this could also be because the residents have seen the plaque many times before.

For me, the most intriguing event that took place when I was there was an encounter with unfamiliar characters such as the old auntie who has lived in the estate for four years and never cared for the plaque although she has noticed it before, until, as she put it, 'you were standing and looking around like Sherlock Holmes like that. I become curious *lah* [*sic*]'. When I told her about the plaque and what I was doing there, she got interested and started telling me about her parents' accounts (told to her) of the difficult experiences faced during the war – food rationing, shortage of basic necessities, cruel Japanese occupiers – when they were living in Chinatown (close to where Hong Lim Complex today), although we never talked about the *Sook Ching*. Two things struck me about this encounter. First, how these memoryscapes, inserted, as it were, into the everyday, may be easily overlooked, thus speaking to a particular limit of such plaques situated in mundane locations albeit in plain sight, and with nothing material but the narratives to link the present with the past (Marshall 2004; Cooke 2000). In fact, if not for the plaque established in 1995, no one would have guessed of this past of the estate. Ironically, while the materiality of the in situ plaque was meant to *presence* this legacy of the surrounding area to present-day Singaporeans, it has been rendered absent by its location outside the city and in a residential area and perhaps the dry text that tells of the experiences of foreign soldiers (no locals were involved in this history as far as I know) with less resonance for Singaporeans.

This raises questions over the efficacy of the plaque. While NHB had wanted to cast the commemorative net wider by materialising these everyday memoryscapes in everyday locations, it would seem that they are not spectacular or substantial enough (without actual relics) to interest passers-by. Even so, what is more interesting is how the materiality of *my own body* traipsing around the plaque lured others (such as the auntie) to be curious about the plaque, endeavouring her to discover the history of her neighbourhood (see Muzaini 2012, 2015). However, for me, the serendipitous encounter with the auntie was much welcome for the light shed on not only more everyday war memories of her parents but also those of local Singaporeans, which have not received much attention in everyday memoryscapes (with the exception of a few such as the one at Hong Lim Complex). While the account of the miscommunication at the Jurong–Kranji Defence Line was, for me, fascinating as one of many missteps that happened during the war (for others, see Murfett et al. 1999), it was the personal stories of the auntie that affected me more. In this case, it was the materiality of the auntie (and her storytelling) that helped to *presence* for me a part of the war history of Singaporeans that has not been given much emphasis even in major war museums on the island, such as the Changi Chapel and Museum (established in 2001) and the Reflections at Bukit Chandu interpretive centre (in 2003), which tended to focus on war experiences of foreigners in the war and, where locals are depicted, these tend to be the combatants.[8]

Moment 3: Sarimbun Landing Site

> At the Sarimbun [Landing Site] plaque, nothing but a plaque amidst a green landscape, I am reminded of my National Service days. What a strange thing to be thinking about. All the training we did in the forests, cemeteries, God knows where ... for Singapore's defence ... I hated it when I was going through it but thinking back, I suppose it was necessary ... we are, after all, vulnerable.
>
> (Fieldnotes, 17 July 2013)

During National Service (NS), Singaporean men are obliged to enter a world of simulated tactical exercises, alongside arduous physical training, to ensure military preparedness should the country ever go to war.[9] Included also is exposure to Singapore's Second World War past as a means of not only learning about battle strategies – what to do and what not – but also to build *esprit de corps* among the men, the war being 'the only campaign that could be remembered ... as a key source of inspiration' (Wong 2001: 234; see Ashplant et al. 2000). In line with this, NS men are taken on excursions to major war museums and cemeteries as part of the Battlefield Tours, where the men (me included, back in the 1990s) are told inspiring stories about the war interwoven with why it was necessary for us to go through NS: because Singapore remains vulnerable; as a small nation, there is a need to ensure that it is always prepared so that it is never again colonised. However, everyday memoryscapes, such as the Sarimbun Landing Site (Figure 3.5), where the Japanese first landed on the island on 8 February 1942, marking the start of the Battle for Singapore, have never been part of these. Located in the jungled area off the beaten track, accessible only by private transport or a walk of about 15 minutes along badly paved roads that are populated by wild dogs, trucks carrying gravel and sand, and signs of 'Protected Area' (marking territory where military trainings still take place), the 'open-book'-shaped plaque is probably only encountered by history and heritage buffs like me or those who use the scouting and adventure camping facilities in the vicinity stumbling upon it by chance.[10]

When I last visited it in 2013, it became one of the most resonant sites for me, particularly in terms of eruptions of serendipitous memories that can occur when they are least expected; not only memory intended but also memory unintended (Crang and Travlou 2001; Hebbert 2005). For one, it provoked personal memories of my own NS days, not only of the jungle training we did and why we did it (as the quote above shows), but also the sense of accomplishment when I ended my stint, and of 'old army buddies' with whom I went through the Singaporean-style rite of passage. These memories were triggered not only by the (albeit brief) text on the plaque about the clash between the 5th/8th Japanese Division (that landed) and the Australian 22nd Brigade (defending the ground), supplemented by what I already knew of the battle, but also the 'natural' environs of the location where I remember once having had to dig my own shellscrape to hold vigil through the night in case simulated 'enemies' (actually my own NS superiors) were to tactically emerge. Thinking about the battle story recounted through the memorial amidst the rural landscape, presumably the way Singapore looked before it became the city it is today – one can still see the coast where the Japanese

Figure 3.5 The plaque marking the Sarimbun Landing Site.

Source: Author.

landed – alongside seeing scouts cheering nearby during a tug-of-war also made me reflect upon how the country has progressed over time.

Most interesting, however, was my encounter with a Malay man who was picking durians in the area when I hitched a ride out to the main road in his van when I got nervous about having to traverse the wild dogs roaming the road. Ahmad (not his real name, in his eighties) used to live in the village in the area before his family was relocated to the HDB flats (mentioned previously) as part of Singapore's modernisation period in the 1970s. When I told him why I was there, he then told me stories of how locals in the area too were terrorised by the Japanese during the war, some of whom became victims of massacres that occurred in the Sarimbun area, even on the main road we were driving. This made me reflect upon two things: first, the selectiveness of the memory depicted on the plaque I just visited, which told of the *foreign* soldiers fighting there during the war but not of the locals. Second, it made me think about the irony that while these plaques are placed in the everyday to allow locals to engage with 'the stuff of history', the everyday war memories of the locals, which occurred in virtually the whole land-scape, what Tyner et al. (2012) refer to as history 'hidden in plain sight', have been sidelined. Indeed, I learnt something new about the area's past, not from the formal plaque itself but from a chance encounter with a man named Ahmad.

Conclusion

> If history is restricted to preserving the image of the past still having a place in the contemporary collective memory, then it retains only what remains of interest to present-day society – that is, very little.
>
> (Halbwachs 1980: 80)

This chapter has concerned 3 of the 14 bronze 'open-book'-shaped plaques folded into the everyday landscapes of Singapore in 1995 to commemorate the fiftieth anniversary of the end of the war (although the insights shed could easily be applied to the other everyday memoryscapes such as those installed by STB, PMB and National Parks Board). Together, these serve to awaken the 'ghosts' [read: 'hidden memories'] that would otherwise lay dormant beneath the progressive city that Singapore has become, providing a sense of chronology to otherwise bland streets and landscapes and transforming them into 'vision[s] of history' (Azaryahu and Foote 2008: 179). Rather than analysing the veracity of the stories told through these memoryscapes, however, the chapter analysed my own feelings and experiences while at these sites. Not meant to be representative of the views of all Singaporeans (particularly since I now live in the Netherlands), the intention is to draw attention to how a reader's interaction with such memoryscapes can extend beyond the depicted past to encapsulate fleeting emotions and encounters that form bonds between the self and the collective, hence the adoption (and value) of autoethnography as a way to tease out 'excessive and transient aspects of living' that other qualitative methods may be hard placed to execute (Lorimer 2005: 83; although see Latham 2003 for other ways to do this).

Indeed, as the three moments discussed above clearly indicate, notwithstanding my critique of these memoryscapes – as being too brief, too dry, too biased, too out-of-the-way – there were times when I was able to appreciate them albeit on a more affective register. First, I felt pride when outsiders (i.e. tourists) read and learnt about the history of the nation, the ways in which Singapore has developed over the years, and for how the Singapore state has sought to ensure its involvement in the war is not forgotten, even as these are 'written' from the perspectives of the present, often at the expense of alternative interpretations of the past (see Hoelscher and Alderman 2004). Second, I was enchanted to be able to 'imagine' the past, by reading the texts on the plaques, which then triggered memories of other accounts of the depicted events read elsewhere, even when there are no longer traces of what used to be. Third, I was intrigued by the ways that these formal memoryscapes triggered for me personal memories, which helped to make these sites more resonant. Finally, encounters with ordinary Singaporeans in these moments also opened up new insights into not only what has been depicted but also local accounts of war that have been excluded, thus *presencing* what has been made absent (Muzaini 2015; Tyner et al. 2012). Collectively, these moments show how these memoryscapes may 'direct people to preferred meanings by the most direct route' (Hoelscher 2008: 4) but also prompt memories and encounters not necessarily related to depicted pasts (see Waterton and Dittmer 2014).

It is important, however, to note that what I experienced may be totally different from the next Singaporean. As 'assemblages', memoryscapes represent the convergence of the agencies of human and non-human materialities – bodies, objects, past knowledge, weather etc. – each relating with one another differently towards affecting certain mental, somatic and emotional effects at specific moments of interaction (Waterton and Dittmer 2014). Thus, people bring their own subjectivities to their engagements, forming own meanings and associations between the past and their lives in the present, making their experiences unique and more resonant (see quote at the start of this section). As such, bonds may be forged between human and memorial even without any relationship to the war or cynicism towards formal processes of memory-making. Given that the process of nation-building can be affective and emotional as much as it is cognitive (Park 2010), it was my engagements with the materiality (and narratives) of the plaques, enmeshed with my own subjectivities (gender, age, ethnicity etc.), positionalities (as former museum curator, as researcher etc.), and prior knowledge of the war, that reacted with each other *in those moments*, and making me feel connected to a country I no longer live in, but that still hold special sentiments. Here, everyday memoryscapes thus serve as 'locus of collective memory in a double sense',

> from above, through architectural order, monuments and symbols, commemorative sites, street names, civic spaces, and historic conservation; and it can express the accumulation of memories from below, through the physical and associative traces left by interweaving patterns of everyday life.
>
> (Hebbert 2005: 592)

Where the former may have failed to make me feel much for the nation, it is the latter that has salvaged for me a sense of closeness to the nation, a reminder I am still Singaporean.

Notes

1 These included the British, Dutch, Americans, Australians, New Zealanders and Indians. While there were also a number of combatants from Malaya, for the most part, locals were largely spectators to 'someone else's war' albeit fought on 'local' ground (Fujitani et al. 2001). There were also local collaborations with the Japanese, thus murkying the delineations of exactly who the 'enemy' was then. These could be some of the factors that may have made it hard for the war to be remembered in the early years of independence.

2 Prior to this, the war was coopted in the 1980s for tourism and to meet demands of foreign commemorators who wanted something to be done to preserve the memories of the war within the country, although not yet employed for purposes of nation-building.

3 In the last national census, the population of Singapore consists of 74.2 per cent Chinese, 10.4 per cent Malays, 10.3 per cent Indians and 5.1 per cent Others (data from Department of Statistics, Singapore, 2015).

4 Although spearheaded by different arms of the state, all formal everyday memoryscapes are marked in collaboration with NHB as custodians of the nation's history and heritage.

5 Alongside formal everyday memoryscapes, there are also the more informal plaques and other markers placed by the people themselves, although these are not emphasised here.

6 More recently emplaced war-related PSM plaques still maintain the 'open-book' shapes although now placed on pedestals rather than on the ground as they were before. Where these locations are also legally marked to be preserved, the red plaques will also be there.

7 When I was reading the text on the plaque, there were loud sounds of crickets, which made me wonder if these were the same sounds heard by the soldiers manning the line then.

8 The lack of local stories in many of the war memoryscapes of non-combatant Singaporeans was what led to the establishment of the Memories at Old Ford Factory museum in 2006 (see Muzaini and Yeoh forthcoming).

9 As part of the National Service (Amendment) Act of 1967, Singaporean male citizens at the age of 18 are compulsorily conscripted for full-time military training for periods of between two and two and a half years.

10 During the visit I also encountered Scout Master Andy (not his real name) who told me that he sometimes uses the plaque to tell his scout recruits about the war although this is 'not standard scout-y spiel'. This is testament to how the plaques not only trigger memories for locals but are also useful for the dissemination of history by locals as well.

References

Ashplant, T.G., Dawson, G. and Roper, M. 2000. The politics of war memory and commemoration: contexts, structures and dynamics. In T.G. Ashplant, G. Dawson and M. Roper (eds) *The Politics of War Memory and Commemoration*. London: Routledge, 3–85.

Atkinson, D. 2007. Kitsch geographies and the everyday spaces of social memory, *Environment and Planning A* 39 (3): 521–540.

Azaryahu, M. and Foote, K.E. 2008. Historical space as narrative medium: on the configuration of spatial narratives of time at historical sites, *GeoJournal* 73 (3): 179–194.

Butler, T. 2006. A walk of art: the potential of the sound walk as practice in cultural geography. *Social and Cultural Geography* 7 (6): 889–908.

Clarke, A. and Waterton, E. 2015. A journey to the heart: affecting engagement at Uluru-Kata Tjuta National Park, *Landscape Research*. DOI: 10.1080/01426397.2014.989965.

Cloke, P., Crang, P. and Goodwin, M. 1999. *Introducing Human Geographies*. London: Arnold Press.

Cooke, S. 2000. Negotiating memory and identity: the Hyde Park Holocaust Memorial, London. *Journal of Historical Geography* 26 (3): 449–465.

Crane, S.A. 1997. Writing the individual back into collective memory. *The American Historical Review* 102 (5): 1372–1385.

Crang, M. and Tolia-Kelly, D.P. 2010. Nation, race and affect: senses and sensibilities at National Heritage sites. *Environment and planning A* 42 (10): 2315–2331.

Crang, M.A. and Travlou, P.S. 2001. The city and topologies of memory. *Environment and Planning D: Society and Space* 19 (2): 161–177.

Dwyer, O.J. and Alderman, Derek H. 2008. *Civil Rights Memorials and the Geography of Memory*. Chicago: The Centre for American Places at Columbia College Chicago.

Fujitani, T., White, G.M. and Yoneyama, L. 2001. Introduction. In T. Fujitani, G.M. White and L. Yoneyama (eds) *Perilous Memories: The Asia-Pacific War(s)*. Durham, NC and London: Duke University Press, 1–32.

Halbwachs, M. 1980. *The Collective Memory.* New York: Harper and Rows Publishers.

Hebbert, M. 2005. The street as locus of collective memory. *Environment and Planning D: Society and Space* 23 (4): 581–596.

Hoelscher, S. 2008. Angels of memory: photography and haunting in Guatemala City. *Geojournal* 73 (3): 195–217.

Hoelscher, S. and Alderman, D.H. 2004. Memory and place: geographies of a critical relationship. *Social and Cultural Geography* 5 (3): 347–355.

Holloway, J. and Kneale, J. 2008. Locating haunting: a ghost-hunter's guide. *Cultural Geographies* 15 (3), 297–312.

Jones, O. 2011. Geography, memory and non-representational geographies. *Geography Compass* 5 (12), 875–885.

Kong, L. and Yeoh, B.S.A. 2003. *The Politics of Landscapes in Singapore: Constructions of 'Nation'.* New York: Syracuse University Press.

Latham, A. 2003. Research, performance, and doing human geography: some reflections on the diary-photograph, diary-interview method. *Environment and Planning A* 35 (11): 1993–2017.

Lorimer, H. 2005. Cultural geography: the busyness of being 'more-than-representational'. *Progress in Human Geography* 29 (1): 83–94.

Marshall, D. 2004. Making sense of remembrance. *Social and Cultural Geography* 5 (1): 37–54.

Murfett, M.H., Miksic, J.N., Farrell, B.P. and Chiang, M.S. 1999. *Between Two Oceans: A Military History of Singapore from First Settlement to Final British Withdrawal.* Oxford and New York: Oxford University Press.

Muzaini, H. 2012. Making memories our own (ways): non-state war remembrances of the Second World War in Perak, Malaysia. In O. Jones and J. Garde-Hansen (eds) *Geography and Memory: Explorations in Identity, Place and Becoming.* Basingstoke: Palgrave Macmillan, 216–233.

Muzaini, H. 2015. On the matter of forgetting and 'memory returns'. *Transactions of the Institute of British Geographer* 40 (1), 102–112.

Muzaini, H. and Yeoh, B.S.A. Forthcoming. *Contested Memoryscapes: The Politics of Second World War Commemoration in Singapore.* London and New York: Routledge.

National Heritage Board (NHB). 2013. *Singapore in World War II: A Heritage Trail,* Singapore: NHB.

Park, H.Y. 2010. Heritage tourism: emotional journeys into nationhood. *Annals of Tourism Research* 37 (1): 116–135.

Ripmeester, M. 2010. Missing memories, missing spaces: the Missing Plaques Project and Toronto's public past. *City, Culture and Society* 1 (4): 185–191.

Samuel, R. 1994. *Theatres of Memory: Past and Present in Contemporary Culture.* London: Verso.

The Straits Times, various issues.

Thrift, N.J. 2000. Non-representational theory. In R.J. Johnston, D. Gregory, G. Pratt and M. Watts (eds) *The Dictionary of Human Geography.* Oxford: Blackwell Publishers, 556.

Till, K.E. 2005. *The New Berlin.* Minneapolis: Minnesota University Press.

Tolia-Kelly, D. 2004. Locating processes of identification: studying the precipitates of re-memory through artefacts in the British Asian home. *Transactions of the Institute of British Geographers* 29 (3): 314–329.

Tyner, J.A., Alvarez, G.B. and Colucci, A.R. 2012. Memory and the everyday landscape of violence in post-genocide Cambodia. *Social and Cultural Geography* 13 (8): 853–871.

Waterton, E. and Dittmer, J. 2014. The museum as assemblage: bringing forth affect at the Australian War Memorial. *Museum Management and Curatorship* 29 (2): 122–139.

Wong, D. 2001. Memory suppression and memory production: the Japanese Occupation of Singapore. In T. Fujitani, G.M. White and L. Yoneyama (eds) *Perilous Memories: The Asia-Pacific War(s)*. Durham, NC and London: Duke University Press, 218–238.

Wylie, J. 2005. A single day's walking: narrating self and landscape on the South West Coast Path, *Transactions of the Institute of British Geographers* 30 (2): 234–247.

Yerushalmi, Y.H. 1989. *Zakhor: Jewish History and Jewish Memory*. New York: Schocken Books.

4 Multiple and contested geographies of memory

Remembering the 1989 Romanian 'revolution'

Craig Young and Duncan Light

This chapter is concerned with the memory and commemoration of the violent events and civil war that were part of the complex downfall of Romanian Communism in 1989–1990. The departure from state-socialism in Romania was distinguished by its violent nature, with the Romanian 'revolution' leaving over a thousand people dead and around 3,500 wounded, with more deaths and injuries in the *Mineriadă* of 1990. What exactly happened during the fall of the Ceauşescu regime in December 1989 is still the subject of debate. Together with the subsequent trajectories of Romanian post-socialist politics, which have been marked by considerable continuities from the Communist regime, this has produced a complex context in which commemoration and memory of these events remain highly contested. Official, state-led processes of commemoration do not resonate with popular memory, which remembers in different ways that ignore or contest the narratives of the post-socialist nation-state.

This case therefore provides an important context in which to explore the multi-faceted nature of the memory of war and violence. The nature of memory formation in post-socialist Romania points to the need to explore the inter-relationships and entanglements between official (state-led) memory and variegated collective and individual memories. There is certainly a materiality to this, as the state and other elites have created particular landscapes encoded with their preferred readings of 1989. However, popular imaginings of these events are expressed in and through different spatial settings and processes, which require analysis of their 'more than textual, multi-sensual worlds' (Lorimer 2005: 83). Memories of conflict are constructed within the inter-relationships of landscape/materiality/object, practice, atmosphere, affect and environments, and how these complexly shape and reshape each other (Thrift 1996, 2004; Lorimer 2005; Anderson 2009; Anderson and Harrison 2010). In this chapter we therefore explore the 'more-than-human' worlds of memories of civil strife using landscape analysis, media and archival sources, fieldwork at memorial sites and focus groups with young Romanians (conducted in 2009 as the twentieth anniversary of the revolution approached) to trace how place, rituals, performances, identities and embodied and affectual experiences intertwine to shape memory.

To begin with we establish, as far as is possible, a short narrative of the events of 1989–1990 associated with the end of the Ceauşescu regime. We then explore the nature of commemorative landscapes, their performance, and emotional and affectual responses to them in two ways: official (state-led) attempts at commemoration; and popular practices and imaginings. Here we do not wish to suggest that these are simply opposites, though they can be distinguished in terms of their spatial arrangements and how they are performed, but to draw attention to how official and public processes of memory formation interact. The analysis therefore considers material landscapes and representations initially, but also pays close attention to the 'more-than-representational' processes of memory, in which personal, emotional and affectual responses, and bodily practices and performances play a key role. To conclude we consider the importance of the locally contingent national context and wider social processes for the nature of this memory formation.

The 1989 Romanian 'revolution'

In 1989 the Socialist Republic of Romania was ruled by Nicolae Ceauşescu, an autocratic and neo-Stalinist dictator. Ceauşescu's decision to pay off Romania's national debt in the 1980s caused unprecedented austerity since most food was exported and energy was rationed at the same time as Ceauşescu pursued a Pharaonic project of rebuilding the capital city. Though Ceauşescu and his wife Elena were universally unpopular, he was the subject of an extravagant personality cult constructing him as the culmination of Romania's historical development and the embodiment of the peoples' aspirations (Tismaneanu 2003).

On 16 December 1989, plans in Timoşoara to evict a dissident Hungarian church minister escalated into mass public protest. On Ceauşescu's orders this was quashed by the army and internal security forces (*Securitate*), leaving 21 dead and over a hundred wounded (Siani-Davies 2005). Ceauşescu called a mass public rally in *Piaţa Palatului* (Palace Square – see Figure 4.1) in Bucharest on 21 December intended to demonstrate his control. However, he was heckled by the crowd and the dictator's confusion was broadcast live on television. Realising that this was their opportunity to bring down Ceauşescu's regime, Romanians took to the streets in protest. The army and *Securitate* opened fire, leaving 49 dead and over 400 wounded but this failed to deter the crowds who, the following day, stormed the Communist Party Headquarters. Ceauşescu escaped by helicopter but was later captured. A group called *Frontul Salvării Naţionale* (National Salvation Front (NSF)), led by Ion Iliescu, took power. The army now declared itself on the side of the people and three days of civil war followed during which the army fought against forces supposedly loyal to Ceauşescu. These events culminated with the trial and execution of the Ceauşescus on 25 December. The NSF committed itself to abolishing the Communist Party and supporting political pluralism and economic reform.

These events, in which 1,104 people died and 3,352 were wounded, became known as the 'Romanian revolution'. Initially it appeared to be a genuine

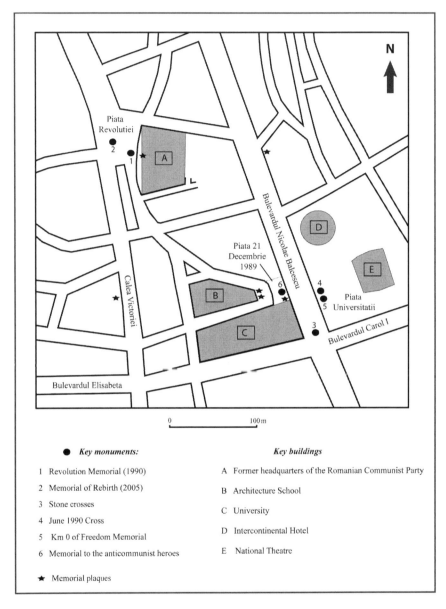

Figure 4.1 Map of locations.

Source: Authors.

'people's revolution', but by January 1990 doubts were being expressed about the 'official' version of events. It was clear that the NSF had organised itself after the revolution began (Sislin 1991) but quickly took control. It was dominated by second-tier Communist Party members and Iliescu himself had once been a high-level Party figure. Critics suggested that the revolution was in fact a *coup d'état* by a faction within the Communist regime intent on overthrowing Ceauşescu without relinquishing power and, furthermore, there are claims that the NSF may have been involved in prolonging the conflict in order to cement its position as the 'guardian of the revolution'. Much of what happened remains shrouded in mystery but Siani-Davies' (1996) account of events – that a revolutionary situation arose without producing a revolutionary outcome – is a good summary. While in official discourse these events are referred to as 'the revolution', most Romanians are sceptical and subscribe instead to the *coup d'état* interpretation, referring instead to a 'so-called' or 'stolen' revolution. Thus throughout this chapter we use the term 'revolution' for the sake of simplicity although we are fully aware of the highly contested nature of this term. Indeed, it is the highly contested nature of these events that are significant in shaping the ways in which they are remembered.

Elite landscapes of memory and popular responses

Studies of memory frequently highlight the importance of the state in shaping remembering or forgetting in constructions of 'the nation' (e.g. Nora 1989; Halbwachs 1992; Legg 2005a; Forest et al. 2004), and nation-states and other elites frequently try to shape popular imaginings by producing particular material landscapes. Thus we begin our analysis with an analysis of elite, particularly state, commemorative landscapes but then look beyond this to the little-understood public consumption of elite landscapes. Here we argue that the state and other elite groups have attempted to sustain a particular memory of the revolution but that the broader public's emotional and affectual responses to elite landscapes and narratives do not cohere around this narrative.

Following Ceauşescu's overthrow, the Romanian state led the way in shaping memorialisation of the revolution (Siani-Davies 2001) and sought to institutionalise the 'peoples' revolution' interpretation of events. The status of 'revolutionary' was bestowed on anyone claiming to have participated in the events granting them privileges, such as tax breaks. Cities that had experienced fighting were declared 'martyr cities' (e.g. Braşov, Sibiu and Târgu Mureş). Between 1989 and 2009, most of the 216 new pieces of legislation relating to the revolution were concerned with honouring 'revolutionaries' (Chamber of Deputies 2011). Other levels of the state were also active. In the early 1990s the street names commission of Bucharest *Primărie* (City Hall) renamed 29 streets in honour of the revolution (Light 2004). Significantly, the emphasis is on the individuals who died rather than the meta-narrative of a popular uprising that overthrew Communism.

This method of commemoration is reflected in a variety of other monuments erected by different parts of the state. During the revolution the NSF took control of the television centre and called upon the people to defend it and 62 people lost their lives there. These events are memorialised in the form of a *troiţa* (a traditional wooden memorial commonly found in rural Transylvania) bearing the inscription: 'The revolution existed in our souls, you made it reality with the price of the supreme sacrifice. Your bravery is recognised! 22 December 1989'. Once again this is a form of memorialisation that focuses on ordinary Romanians rather than elites. Nearby houses still display bullet marks from the fighting. This theme is also apparent in Paul Neagu's sculpture *Century Cross*, erected in 1997 as the result of a competition held by the *Primărie* to produce a monument to the victims of the revolution.

In Bellu Cemetery victims of the revolution are buried in rows of uniform graves (recalling practices of the burial of war dead) bearing the Romanian tricolour. The metro station opposite the cemetery has been named *Eroii Revoluției* (Heroes of the Revolution). However, this official commemorative landscape is also overlain by personal acts of remembrance, as the graves are decorated with items left by relatives and they act as the sites for personal acts of remembering. Beside the cemetery is the *Biserica eroilor martiri ai revoluției din decembrie 1989* (the Church of the Martyr Heroes of the December 1989 Revolution, 2003) administered by the Romanian Orthodox Church, but built at the initiative of an association of the relatives of the dead. The Romanian Orthodox Church holds special ceremonies each year to commemorate the victims of the revolution at the Patriarchal Cathedral and other significant churches, and these ceremonies are attended by many ordinary Romanians. Thus, official and personal landscapes and practices of remembrance intersect in these multiple spaces of memory in ways that extend and blur elite memory formations beyond the spaces and narratives of the state.

Another elite actor in the commemoration of the revolution is the Romanian Army. Although initially implicated in the repression in Timișoara, the army played a pivotal role in the revolution by switching sides and has thus subsequently positioned itself as the guardian of the ideals of the revolution. The NSF was keen to retain the army's support and so, in January 1991, released funding to the Ministry of Defence to commemorate fallen military personnel (*Monitorul Oficial* 1991). One key memorial is located at Bucharest's international airport where 49 soldiers were killed in a 'friendly-fire' incident. This marble monument features individual memorials to each soldier and a central memorial bearing the inscription: 'To our country speak this truth: we were where we were told to be'. This memorial commemorates those who died as a result of a tragic mistake that occurred within the chaos of the revolution, rather than the overthrow of the Ceaușescu regime. The Army also maintains a gallery in the National Military Museum including a memorial inscribed with the names of the soldiers who died, and regular services of remembrance are held here. Displays show uniforms, military paraphernalia from December 1989, a 'revolutionary' Romanian flag (with the emblem of the Communist republic cut out of its centre) and photographs from the revolution with captions such as 'In defence of the Revolution, Bucharest 1989' and 'The army is with us', representing the role of the army as supporters of the revolution.

Thus, multiple elites have actively shaped a complex set of memorial practices. However, it is the state that has been the most powerful actor involved in shaping memorial landscapes in Bucharest, and we now turn to its efforts and popular responses to them. The NSF swiftly set about creating an official commemorative landscape. *Piata Palatului* (Palace Square) was renamed *Piaţa Revoluţiei* (Revolution Square) (*România liberă* 10 January 1990), an attempt by the new regime to sanctify the place where the revolution began. As Azaryahu (1996, 2009) has argued, in a revolutionary context the renaming of urban places is a powerful political statement that something has changed, and a way for a new regime to declare its aspirations. A decree of 7 February 1990 declared Revolution Square to be a '*loc al spiritului dezinteresat, care va iesi din sfera pasiunilor politice*' (*Monitorul Oficial* 1992). This translates approximately as 'a place of the disinterested spirit, beyond the sphere of political interests', and is thus an attempt to depoliticise the 1989 events and to enshrine the official narrative of the 'people's revolution', closing down other memories.

The first official memorial was inaugurated in Revolution Square in 1990 on the initial anniversary of the revolution. This pyramid-shaped memorial bears the text 'Glory to Our Martyrs', but the monument is small-scale and sober rather than triumphant. The apex points to the balcony where Ceauşescu gave his last speech – the event that everyone can agree upon as the start of the revolution. The flowers in the monument may represent a stream of blood, again affirming the narrative of the bloody revolution in which the people rose to overthrow a tyrant.

Thus *Piaţa Revoluţiei* has become a site in which successive post-Ceauşescu governments have commemorated the revolution as the event that brought about the overthrow of Ceauşescu. The square has come to be associated with opposition to (and the end of) Communist Party rule (an inversion of its Communist-era meanings when it was a place for Party ceremonies). In 2003, the Social Democratic Party government (the heir to the NSF with Iliescu again the state president) erected a second memorial here. As Failler (2009) suggests, state strategies of remembrance at official sites can be attempts to shield the state itself from reflecting upon its own role in traumatic events. The monument is known as the *Memorialul Renasterii* (the Memorial of Rebirth – see Figure 4.2) and the winning design was chosen by Iliescu himself. It is a much larger monument than that erected in 1990, and the memorial complex includes a short 'Avenue of Victory' leading to a 'Square of Reflection', a 'Wall of Memory' and the centrepiece of the ensemble, a 'Pyramid of Victory'. Inaugurated in August 2005, it is the central location for official ceremonies commemorating the revolution.

However, analysis of the reception of this monument by varying elites and publics illustrates how official memoryscapes do not simply shape collective memories but can be open to different readings, subversion and/or resistance (Atkinson 2008; Legg 2005b; Forest et al. 2004; Foote and Azaryahu 2007). This was a top-down project, led by the state elite, which enjoys little popular support. The Romanian press dismissed the monument, saying it was built with taxpayers' money without public consultation. Architects and critics derided it as of little architectural merit (*Adevărul* 2005a). Organisations representing those who participated in the revolution have also rejected the monument (*Adevărul* 2005b). In fact, it is so unpopular that most of the mayoral candidates in 2008 promised to demolish it if elected (*Gândul*, 2008).

Figure 4.2 The 'Memorial of Rebirth' in *Piaţa Revoluţiei* (aka the 'Impaled Potato').

Source: Authors.

This negative response to this official memorial landscape is also reproduced more widely in the everyday personal practices of the Romanian public. In everyday discourse the monument is referred to as the *ţepuşa cu cartoful* ('the stake with the potato'), or 'the impaled potato', or even just 'the potato' (see Figure 4.2). The indifference towards this memorial mirrors broader scepticism about the official narrative of the revolution and that the 'potato' is regarded as the project of a government with roots in the Communist regime. This narrative parallels the argument of Forest et al. (2004) that in post-socialist regimes the public may associate state-led commemorative practices with continuity in state power from the Communist period. In terms of embodied practice, on the one hand, the monument is largely ignored by Bucharesters who rarely enter this official memorial space, even during state-led commemorative events. On the other hand, the monument is occasionally the subject of mocking graffiti. For example, stencilled graffiti around the monument state 'Monumentul erorilor', which plays on the similarity in Romanian between the words for 'heroes' and 'errors': thus 'the monument of errors' rather than 'the monument of heroes'.

Given this lack of engagement with the official site of memory, it is people's unseen relationships with the state-led memorial material landscape that is significant for understanding the nature of memory formation in response to them. The state-sponsored memorials intend to construct *Piaţa Revoluţei* emotionally and affectually as a solemn place of remembrance. However, few Bucharesters regard the square in this way, dismissing it as 'just a stage' and rejecting the meanings intended by the state. Here a focus on affectual, emotional and sensational responses (Thrift 2007) to the monument is significant. As one respondent stated:

It has no meaning for me. I really don't like the monument. It's a joke I have with my friends, but the only thing about it that makes me sad or uncomfortable is the marble benches [i.e. trying to sit on them].

(Focus group, female, twenties)

Another stated that: 'There's nothing about it that is about respect. I don't feel sorrow . . . it makes me angry, actually' (female, twenties). While there are personal emotional and affectual responses to the monument – apathy, dislike, anger – they are very much not those intended by the state (solemnity, celebrating overthrowing Communism).

In this light the focus group discussed what would form a suitable memorial for the revolution, with one saying that: 'I think the revolution should have a common monument that represents what people feel like, not some statue designed by an architect. Like a space where you can express yourself as a citizen' (female, twenties). Here, a suitable memorial space that would allow citizens to practice their own senses of commemoration would be a very different space.

Significantly, the official commemorative spaces of the revolution have not become the focus of a shared sense of public memory. Instead, they are largely treated with indifference or even explicitly rejected, illustrating what Rose-Redwood (2008) terms the performative limits of official commemorative landscapes. While elites seek to inscribe the public landscape in different ways, their efforts can be ignored or contested by citizens through myriad counter-performances of memorial spaces and the desired emotional politics are not achieved. This, then, raises questions about 'alternative' memories of the revolution embedded within different landscapes and popular practices.

Alternative commemorative practices and landscapes

Many Romanians are indifferent about the official narrative of the revolution, but agree that the individuals who died deserve remembrance, thus producing a separate memorial landscape focusing not on the 'big events' – the downfall of Communism – but instead remembering those killed. Here we explore the alternative geography of these counter-memorials (Young 1992) and the differing personal performances and emotional responses underpinning these countermemories (Legg 2005a; Tabar 2007; Zerubavel 1995; Alderman 2010).

The principal site of countermemory is *Piața Universității* (University Square – see Figure 4.1). This square – and specifically a small, previously unnamed area within it now called *Piața 21 Decembrie 1989* – has special significance. A number of people died here during the first night of the revolution and it quickly became a site of informal remembrance as Bucharesters brought wooden crosses and candles and erected improvised, ephemeral memorials (Beck 1993). It contains small and unobtrusive memorial plaques and crosses with simple inscriptions such as 'For the heroes of the Revolution, 21–22 December 1989' and 'Here they died for freedom, 21–22 December 1989'. There is nothing to identify who erected them – they represent informal, private commemorations.

In *Bulevardul Bălcescu*, which runs past *Piața 21 Decembrie 1989*, are a number of much older stone crosses, one of which has been inscribed by a local painter with the text 'To the heroes of the revolution'. A wooden cross was also erected here by a group representing revolutionaries with the blessing of the Romanian Orthodox Church. Further along the boulevard is a collection of memorials to the first victims of the revolution (Figures 4.1 and 4.3). Again, these are small, private memorials, grouped informally on the edge of a busy pavement beside a shop entrance, some of which were erected by relatives. These sites also demonstrate the performative nature of memory as they are frequently decorated with candles and flowers. On the Post Office building a plaque erected by Post Office workers to commemorate a colleague reads 'On 25 December 1989, on the fourth day of the Romanian revolution our innocent colleague was taken at only 22 years old'. All of these acts of commemoration are initiated outside the state and they inscribe a parallel set of countermemories onto the landscape. They are deliberately placed in different locations from official monuments and focus on the individuals who died rather than the revolution itself. These are deathscapes (Maddrell and Sidaway 2010) or sites of death, loss and mourning (Hallam and Hockey 2001) in which private grief is publically displayed through smaller, individual and highly personalised forms of memorialisation. Acts such as lighting candles and laying flowers constitute a form of 'commemorative vigilance' (Nora 1989) performed by ordinary Bucharesters, acts that involve a more sensual and tactile engagement with sustaining memory than the efforts of state institutions, which are largely centred on conventional monumental structures.

However, *Piața Universității* was also the site of further events that reinforce these countermemories. In 1990 it was apparent that the NSF was dominated by former Communists. When it announced its intention to stand in the 1990 elections, students and young people set up a protest camp in *Piața Universității*. The NSF won convincingly and Iliescu was elected president with 85 per cent of the vote provoking further protest. In response, thousands of miners were brought to Bucharest on chartered trains and told that Romania's new democracy was under attack from anarchists, deviants and foreign agents. On reaching University Square they attacked the protesters and ransacked the university, with most of the brutal violence occurring in *Piața 21 Decembrie 1989*. Officially seven people died but the actual total is believed to be in the hundreds. This shocking event (the *Mineriadă* or 'Miners' Rage') demonstrated that the post-Ceaușescu regime was as willing as its predecessor to use violence.

Unsurprisingly, the state has made no attempt to commemorate these events. However, the square has become the most significant site of countermemory in the capital and it contains a diverse range of memorials to the young people killed in the *Mineriadă*. The university building in *Piața 21 Decembrie 1989* bears a memorial plaque with the inscription 'Here students and lecturers fought for freedom and civil rights in December 1989 and April–June 1990'. The Architecture School opposite was also the site of spontaneous and personal commemoration as it was extensively graffitied with protest slogans throughout the 1990s (this was cleared in 2001 when the Social Democratic Party, successor to the NSF, was in

Figure 4.3 Group of personal memorials on Bălcescu Boulevard: top left – 'In this spot
true heroes died, 21 December 1989'; top right – plaque commemorating
Luiza Mioara Iancu (erected by the family 10 years after the revolution);
bottom left – plaque commemorating Mihai Gîtlan, the first person to die in
the revolution along with 13 others, apparently while kneeling and praying
(erected a year after the event by the family); bottom right – informal metal
cross inscribed 'In memory of the young people who fell in the revolution'.

Source: Authors.

power). In the centre of *Piaţa 21 Decembrie 1989* is a metal cross erected by a local artist, Constantin Popescu. Writing on the cross states that it is 'for the anti-communist heroes' and invites passers-by to place a flower in memory of those who died, a practice that is entirely non-state led. The cross is regularly cared for and repainted, apparently by the painter himself.

Piaţa Universităţii – and in particular *Piaţa 21 Decembrie 1989* – is an informal but powerful site of countermemory that directly challenges and subverts official narratives, reminding ordinary Bucharesters that the deaths of December 1989 did not bring about the desired political change (Antonovici 2009). What happened in June 1990 further challenged the official state narrative of a 'people's revolution'. This is a place that has far more emotional resonance for the people of Bucharest than *Piaţa Revoluţiei*. As one focus-group participant stated:

> For me the area around University Square is more powerful . . . [The graffiti on the Architecture School building] should have been left. That would have been a better monument than the 'impaled potato'. So a better monument would not even need to be a monument. There are bullet holes in the walls of the National Museum of Art. That is a better monument.

> (Male, twenties)

In the popular imagination there is a distinction between the official commemorative landscape of *Piaţa Revoluţiei* and the informal memoryscape of *Piaţa Universităţii*. The latter is inseparable in public memory from resistance to Communism and post-1989 neo-Communism. This is reflected in people's embodied performance of these spaces. In contrast to the formal, disciplined and staged ceremonies that unfold in *Piaţa Revoluţiei*, in *Piaţa Universităţii* people perform memory in a much more mundane, everyday and informal way (such as laying flowers). As a result the affectual atmospheres of remembrance are entirely different in each space, representing a diversity of post-conflict landscapes of affect (Navaro-Yashin 2009, 2012).

The post-Ceauşescu state has always had an ambivalent relationship with *Piata Universităţii*, particularly when former Communists were in power. However, apart from removing graffiti, there have been no attempts by the state to reinscribe the meaning of the square or to intervene with these alternative, personal acts of commemoration. This tension between official and popular memory was apparent during the twentieth anniversary of the revolution in 2009 when official ceremonies unfolded in *Piaţa Revoluţiei*, but it was in *Piata Universităţii* that revolutionaries and Bucharesters gathered to remember.

Countermemory beyond landscapes and sites of commemoration

Countermemories of the revolution are also formulated beyond both official and personal commemorative landscapes, illustrating the wider production of memory throughout society in more quotidian and mundane spaces (Atkinson 2007, 2008).

Here, memory is psychological and embodied (Legg 2005a), and commemoration is an individual process within a broader social construction of memory (Boym 2001). In this section we explore the nature of these countermemories not only as personal memories, but also how they are shaped by wider social processes within different spaces such as the family or mediatised and virtual spaces.

Within the focus groups, when asked if their families would do anything special to mark the anniversary of the revolution, one respondent highlighted this lack of engagement with official processes, saying: 'Not really. *They know*. They don't have to do anything special' (male, twenties). Memory is thus formed and sustained through internalised and often individualised practices: as this respondent said: 'We might light a candle, or something.' This point was also discussed by several of the focus-group respondents (emphasis added):

> I wasn't involved in the revolution, but you are aware, but *in the back of your mind*, because you know that if it wasn't for the revolution we wouldn't be living as we do now, but it doesn't matter in the way that you would celebrate it. *It's personal, it's inside us* [general agreement]. I see it in a very symbolic way, the revolution means a step towards freedom, towards being what we are today.
>
> (Male, twenties)

Another commented on how it was important that their memories of the revolution functioned in this way:

> I don't like to see it on the news. It makes me sick if there's any cynicism about it. If I saw any sign of it being 'commercialised', that would be it. *It's something that's personal, that's inside me, and that's how I think about it.*
>
> (Female, twenties)

Here there is an explicit rejection of any attempt by the state or media to take control of and shape narratives of the revolution. Thus memories of the revolution are more the product of a population shaping their individual memories rather than the outcome of state-led processes of memory formation. However, memory is differentiated among different generations. Some young people retain direct memories of things that happened during December 1989, though at the time they were shaped by interaction with their families and, of course, were seen from the perspective of a child. For example:

> At the time I was in Bucharest with my parents. I remember my mum moving the furniture onto the window ledge [to stop bullets] and crying because my father was out on the streets. (Female, twenties)

> We were living in Ploieşti. My parents were watching TV and it stopped and we realised that the revolution was really happening. My parents were happy but scared. (Male, twenties)

However, there is a generation born after the revolution that does not have any direct memories but is nevertheless able to narrate post-memories or re-memories (Hirsch 1996; Tolia-Kelly 2004), which illustrate the trans-generational shaping of memory (Adelman 1995; Weingarten 2004). As the focus group discussed their memories of the revolution, it became obvious that they were shaped by different forms of interactions within different families:

> My parents don't have a problem. If I ask a question they answer it. But it is not a topic which comes up a lot. My neighbours used to discuss it, how it wasn't a Christian thing to shoot [the Ceauşescus] on Christmas Day. (Female, twenties)

> My grandfather always talks about it, he is passionate about history. (Female, twenties)

The transgenerational construction of individual memory and re-/post-memory thus varies considerably in relation to family dynamics, with different family members exhibiting different degrees of willingness to remember, or forget.

These re-/post-memories among younger people are also highly mediatised (van Dijck 2007, 2008; LeMahieu 2011). Younger people 'encounter' the revolution through the education system, television, documentaries and books in ways that shape their understandings and memories:

> I remember after the Communist period there were many documentaries about the revolution. And there are many books. And now they are trying to write the truth about it. (Male, twenties)

> Usually in December there's something on TV around that time. And they say 'this happened' and they retell that story. (Female, twenties)

However, younger generations' memories are additionally mediated by the development of new technologies, especially the internet. As van Dijck (2008: 70) argues 'embedded in networked systems, pictorial memory is forever distributed, perpetually stored in the endless maze of virtual life'. Virtually distributed representations of the revolution are key in shaping younger people's memories: 'And of course there's YouTube and whoever wants to can see those movies and Ceauşescu's last speech' (Male, twenties). There are many internet versions of Ceauşescu's last speech and the crowd turning on him, which have received millions of views.

The mediated nature of memories among this generation was strikingly illustrated when the focus-group members unanimously claimed to remember the execution of the Ceauşescus, even though they probably never actually saw it. The execution was shown repeatedly on television (to demonstrate the Ceauşescus really were dead) and the footage can be viewed on internet sites such as YouTube. The graphic violence of their execution is ingrained into young peoples' memories because of its traumatic content, a form of memory that is sometimes referred to as 'flashbulb memories' (Kensinger and Schacter 2006):

[W]hen they shot the Ceauşescus, I still have that 'screenshot' in my mind. I don't know if I saw it later or I still have it in my mind from that period. My parents were discussing if it is normal or not to kill him on Christmas Day. (Female, twenties)

I remember the video of the shooting. They were scared. And they shot them even after they were dead. They kept shooting. It was horrible. (Female, twenties)

However, these responses indicate that, although they hold strong memories of this event, it was difficult for these younger people to untangle them from subsequent media exposure, complicating the relationship between media and memory. The focus group was asked 'did you see the image of the shooting on TV': '[general confusion among the group] . . . I don't know if I saw it at the time. I don't know if it was broadcast on TV. They were lying on the floor with the blood flowing, and their eyes open' (female, twenties).

Thus there is not a simple linear relationship between distributed digital representations and memory formation. Van Dijck (2008) argues that digital technology loosens the control of individuals (and, by extension, elites) to maintain power over how images are manipulated and framed in public contexts such as the internet, a process heightened by the democratisation of memory (Atkinson 2008) in the post-socialist context. Digital music and video technologies allow younger generations to mutate iconic images into new representations of the past, which inform their memories in different ways. As one focus-group respondent said: 'There's even a music remix using Ceauşescu's last speech. It's art . . . it's turning into art' (male, twenties). For example, a remix by DJ RA7OR samples part of the televised speech where Ceauşescu repeatedly shouts to the crowd 'Alo! Tovarăşi! Aşezaţi-vă liniştiţi!' ('Hello! Comrades! Everyone be quiet!'). The iconic image of Ceauşescu's last speech has been reworked within youth culture in a way that frames the remembrance of the revolution quite differently, which illustrates that 'memory products . . . invite subversion or parody, alternative or unconventional enunciations' (van Dijck 2007: 7).

Thus, for most Romanians, the formation of memories of the revolution does not occur through official commemorative landscapes but instead is personal and shaped in the quotidian space of the home through mundane interactions with their families, and during leisure activities where they interact with mediatised memory products, which are consumed and reproduced in complex ways. Here memory is less bounded and more fluid and continually reconstituted (Atkinson 2008).

The existence of these diverse individualised, institutionalised, transgenerational and mediated countermemories has important implications for how personal and public memory formation interacts with official attempts to commemorate the revolution as the foundation myth for a 'new' Romania (Boia 2002). In official discourse the revolution is the event that symbolises the end of Communism. In this sense the state has attempted to mobilise remembrance of the revolution within nation-building strategies as part of creating an 'imagined community' (Anderson 2006). However, countermemories of the revolution work in a very different way – the revolution is remembered in terms of respect and sympathy for those individuals who died but with few specifically national resonances. For example, a focus-group respondent said: 'You celebrate the National Day because it's something that you understand and everybody does it, but the revolution doesn't have that meaning' (female, twenties).

Thus the celebration by the state of a national day has resonance but attempts to use memories of the revolution for this purpose do not work. Another focus-group respondent illustrated this when he gestured to the pub sound system playing the Michael Jackson song 'Thriller' and said: '[L]ike, it's more important that Michael Jackson died [general laughter] . . . yeah, actually it is, that's something everyone talks about . . . if the media made the revolution more 'trivial' then maybe everyone would talk about it too' (Male, twenties).

The contested nature of the revolution, and the way that it is remembered by Romanians of all ages, means that it is difficult for the state to generate and sustain a coherent memory of the revolution to underpin new myths of '*the* nation'. Instead, people's everyday practices and popular traditions (Edensor 2002) do not reinforce the state's narrative. These various sites of countermemory thus function to 'rebut the memory schema of a dominant class' (Legg 2005a: 181).

Conclusion

As Alderman (2010: 90) notes, places of memory narrate history through selective processes of remembering and forgetting, leaving 'what is defined as memorable as . . . open to social control, contest and renegotiation'. In this chapter we have explored how different elites and publics have engaged in the 'work' of memory in quite different ways, which demand an understanding based on interrogating the entanglement of material landscapes and objects, embodied performances and rituals, atmospheres and emotional and affective responses. In the Romanian case, these entanglements were shaped by the contingencies of local and national events and social and political changes.

Post-socialist memory formation is shaped by people's experience of national variants of state-socialism and one legacy is a mistrust of the 'public' sphere and politics (Nadkarni 2012; Gille 2012). Decades of state-sponsored attempts to shape memory in support of political projects have produced post-socialist societies that may feel little affiliation with elite visions of the nation and this feeling maybe stronger in societies with more traumatic experiences of Communism. Furthermore, post-socialist societies may question attempts by new regimes to suggest a 'clean break' with the past if they perceive elements of continuity rather than change in political power structures (Forest and Johnson 2002; Gille 2012). In the Romanian context, Popescu-Sandu (2012) suggests that such a situation provokes a 'paralysis' in politicising memory and follows Boia's (2002) notion of a 'mythological blockage' of proper consideration of Romania's past.

Romania's exit from state-socialism in 1989 was highly complex politically and – unusually for the former Eastern Europe – was marked by violent conflict. Memories of these events have been shaped by strong political continuities from the Communist period, combined with the rapid adoption among Romanian citizens of the 'stolen revolution' version of events, reinforced by the subsequent violence of the *Mineriadă*. In this context, memories of the revolution have developed in a very particular way, with popular memories being associated with an alternative and separate geography of sites and landscapes, a rejection of official narratives and landscapes, and a subsequent formation of memory in a variety of different places beyond the reach of the disciplining state. Rather than seeing these as two opposing sets of memories, however, it is important to see them as alternative strategies of memory that exist because of how they work against each other.

There is little doubt that the Romanian state and allied institutions (such as the Orthodox Church) will continue to commemorate the revolution, which in turn will retain its importance as the fundamental element of the official foundation myth of the post-Communist Romanian nation. However, there are other drivers of post-Communist Romanian identity, not least integration with Europe. As for the alternative and counter-memories sustained among the general population, these will be sustained by an increasingly older generation and may not be taken up by younger generations, though young people (students in particular) still maintain recognition and pride in those events, and whenever there is a protest young people head to *Piata Universității*, which is becoming a site of protest more than a site of memory. The interaction of the affective qualities of public space and memory is a constantly evolving process.

References

Adevărul. 2005a. Țepușa cu cartof, o nouă statuie înfiptă în asfaltul Capitalei, 20 June, 1.
Adevărul. 2005b. Cartoful din Piața Revoluției – o țeapă de 56 miliarde de lei trasă de Răzvan Theodorescu, 6 July, 16.
Adelman, A. 1995. Traumatic memory and the intergenerational transmission of Holocaust narratives. *Psychoanalytic Study of the Child* 50: 343–367.
Alderman, D.H. 2010. Surrogation and the politics of remembering slavery in Savannah, Georgia (USA). *Journal of Historical Geography* 36: 90–101.
Anderson, B. 2006. *Imagined Communities: Reflections on the Origin and Spread on Nationalism*. London: Verso.
Anderson, B. 2009. Affective atmospheres. *Emotion, Space and Society* 2 (2): 77–81.
Anderson, B. and Harrison, P. 2010. The promise of non-representational theories. In P. Harrison and B. Anderson (eds) *Taking-Place: Non-Representational Theories and Geography*. Farnham: Ashgate, 1–36.
Antonovici, V. 2009. Piața Universității – loc memorial? *Sfera Politicii* 17: 94–99.
Atkinson, D. 2007. Kitsch geographies and the everyday spaces of social memory. *Environment and Planning A* 39: 521–540.
Atkinson, D. 2008. The heritage of mundane places. In B.J. Graham and P. Howard (eds) *The Ashgate Research Companion to Heritage and Identity*. Aldershot: Ashgate, 381–396.

Azaryahu, M. 1996. The power of commemorative street names. *Environment and Planning D: Society and Space* 14: 311–330.

Azaryahu, M. 2009. Naming the past: the significance of commemorative street names. In L.D. Berg and J. Vuolteenaho (eds) *Critical Toponymies: The Contested Politics of Place Naming*. Farnham: Ashgate, 53–67.

Beck, S. 1993. The struggle for space and the development of civil society in Romania, June 1990. In H.G. DeSoto and D. G. Anderson (eds) *The Curtain Rises: Rethinking Culture, Ideology and the State in Eastern Europe*. Atlantic Highlands, NJ: Humanities Press, 232–265.

Boia, L. 2002. *Istorie și mit în conștiința românească* (3rd edition). București: Humanitas.

Boym, S. 2001. *The Future of Nostalgia*. New York: Basic Books.

Chamber of Deputies. 2011. Database of Romanian legislation, www.cdep.ro, last accessed 8 January 2016.

Edensor, T. 2002. *National Identity, Popular Culture and Everyday Life*. London: Berg.

Failler, A. 2009. Remembering the Air India disaster: memorial and counter-memorial. *Review of Education, Pedagogy, and Cultural Studies* 31: 150–176.

Foote, K.E. and Azaryahu, M. 2007. Toward a geography of memory: geographical dimensions of public memory and commemoration. *Journal of Political and Military Sociology* 35: 125–144.

Forest, B. and Johnson, J. 2002. Unravelling the threads of history: Soviet-era monuments and post-Soviet national identity in Moscow. *Annals of the Association of American Geographers* 92: 524–547.

Forest, B., Johnson, J. and Till, K. 2004. Post-totalitarian national identity: public memory in Germany and Russia. *Social & Cultural Geography*, 5: 357–380.

Gândul, 2008. 'Țeapa' din Piața Revoluției sare în ochii candidaților, 21 May, www.gandul.info/politica/teapa-din-piata-revolutiei-sare-in-ochii-candidatilor-2650860, last accessed 5 November 2010.

Gille, Z. 2012. Postscript. In M. Todorova and Z. Gille (eds) *Post-Communist Nostalgia*. Oxford: Bergahn Books, 278–289.

Halbwachs, M. 1992. *On Collective Memory*. Chicago: University of Chicago Press.

Hallam, E. and Hockey, J. 2001. *Death, Memory and Material Culture*. Oxford: Berg.

Hirsch, M. 1996. Past lives: postmemories in exile. *Poetics Today* 17: 659–686.

Kensinger, E.A. and Schacter, D.L. 2006. Reality monitoring and memory distortion: effects of negative, arousing content. *Memory & Cognition* 34 (2): 251–260.

Legg, S. 2005a. Sites of counter-memory: the refusal to forget and the nationalist struggle in Colonial Delhi. *Historical Geography* 33: 180–201.

Legg, S. 2005b. Contesting and surviving memory: space, nation, and nostalgia in *Les Lieux de Mémoire*. *Environment and Planning D: Society and Space* 23: 481–504.

LeMahieu, D.L. 2011. Digital memory, moving images and the absorption of historical experience. *Film & History* 41: 82–106.

Light, D. 2004. Street names in Bucharest 1990–1997: exploring the modern historical geographies of post-socialist change. *Journal of Historical Geography* 30: 154–172.

Lorimer, H. 2005. Cultural geography: the busyness of being 'more-than representational'. *Progress in Human Geography* 29 (1): 83–94.

Maddrell, A. and Sidaway, J.D. 2010 (eds) *Deathscapes. Spaces for Death, Dying, Mourning and Remembering*. Farnham: Ashgate.

Monitorul Oficial. 1991. Hotărâre nr 3 din 5 ianuarie 1991 privind finanțarea și realizarea de catre Ministerul Apararii Nationale, a unor lucrari comemorative pentru cinstirea eroilor cazuti in Revolutia din Decembrie 1989, Nr 7 (15 January).

Monitorul Oficial. 1992. Hotarire nr. 125 din 7 Februarie 1990 privind unele masuri pentru marcarea memoriei eroilor Revolutiei din 22 Decembrie 1989, Nr 148, 30 June, www. cdep.ro/pls/legis/legis_pck.frame, last accessed 9 March 2010.

Nadkarni, M. 2012. 'But it's ours': nostalgia and the politics of authenticity in post-socialist Hungary. In M. Todorova and Z. Gille (eds) *Post-Communist Nostalgia.* Oxford: Bergahn Books, 190–214.

Navaro-Yashin, Y. 2009. Affective spaces, melancholic objects: ruination and the production of anthropological knowledge. *Journal of the Royal Anthropological Institute* NS 15: 1–18.

Navaro-Yashin, Y. 2012. *The Make-Believe Space: Affective Geography in a Post-War Polity*. London: Duke University Press.

Nora, P. 1989. Between memory and history: les lieux de mémoire. *Representations* 26: 7–25.

Popescu-Sandu, O. 2012. 'Let's all freeze up until 2100 or so': nostalgic directions in post-Communist Romania. In M. Todorova and Z. Gille (eds) *Post-Communist Nostalgia*. Oxford: Bergahn Books, 113–128.

România liberă 1990. Pentru comemorarea civilor revoluției, 10 January, p. 10.

Rose-Redwood, R. 2008. 'Sixth Avenue is now a memory': regimes of spatial inscription and the performative limits of the official city-text. *Political Geography* 27: 875–894.

Siani-Davies, P. 1996. Romanian revolution or coup d'état? A theoretical view of the events of December 1989. *Communist and Post-Communist Studies* 29: 453–465.

Siani-Davies, P. 2001. The revolution after the revolution. In D. Light and D. Phinnemore (eds) *Post-Communist Romania: Coming to Terms with Transition*. Basingstoke: Palgrave, 15–34.

Siani-Davies, P. 2005. *The Romanian Revolution of December 1989*. Ithaca, NY: Cornell University Press.

Sislin, J. 1991. Revolution betrayed? Romania and the National Salvation Front. *Studies in Comparative Communism* XXIV: 395–411.

Tabar, L. 2007. Memory, agency, counter-narrative: testimonies from Jenin refugee camp. *Critical Arts* 21: 6–31.

Thrift, N. 1996. *Spatial Formations*. London: Sage.

Thrift, N. 2004. Intensities of feeling: towards a spatial politics of affect. *Geografiska Annaler* 86 (B): 57–78.

Thrift, N. 2007. *Non-Representational Theory: Space, Politics, Affect*. London: Routledge.

Tismaneanu, V. 2003. *Stalinism for All Seasons: A Political History of Romanian Communism*. Berkeley: University of California Press.

Tolia-Kelly, D. 2004. Locating processes of identification: studying the precipitates of re-memory through artefacts in the British Asian home. *Transactions of the Institute of British Geographers* 29: 314–329.

van Dijck, J. 2007. *Mediated Memories in the Digital Age*. Stanford, CA: Stanford University Press.

van Dijck, J. 2008. Digital photography: communication, identity, memory. *Visual Communication* 7: 57–76.

Weingarten, K. 2004. Witnessing the effects of political violence in families: mechanisms of intergenerational transmission of trauma and clinical interventions. *Journal of Marital and Family Therapy* 30: 45–59.

Young, J.E. 1992. The counter-monument: memory against itself in Germany today. *Critical Inquiry* 18: 267–296.

Zerubavel, Y. 1995. The multivocality of a national myth: memory and counter-memories of Masada. *Israel Affairs* 1: 110–128.

5 Wrecks to relics

Battle remains and the formation of a battlescape, Sha'ar HaGai, Israel

Maoz Azaryahu

National memory features prominently in the geopolitics of national identity. As is often the case, the question 'who we are' is often answered in terms of the answer to the question 'where do we come from?'. Potentially contested and susceptible to revisions, national memory represents an ongoing process aimed at the production of historical consciousness. Notably, national memory is culturally manifest in the form of a space–time matrix of commemorations that reproduce and introduce history as a contemporary cultural experience (Foote and Azaryahu 2007). In particular, the geography of memory locates history and its representations in space and landscape. As Nuala Johnson observed, 'space or more particularly territory is as intrinsic to memory as historical consciousness in the definition of national identity' (1995: 55). Linking together history, memory and territory is essential for the conceptualization of a land as a homeland. No wonder, then, that landscapes and sites that cast a certain version of history into a mold of commemorative permanence belong to the symbolic foundations of modern nationhood (Smith 1991). When associated with historical events and pertaining to national memory, such landscapes and sites conflate historical events and contemporary sights, and in this semiotic capacity substantiate the nation in space (territory) and in time (history and memory).

Fought against Arab-Palestinian irregulars and Arab armies, Israel's War of Independence (December 1947–January 1949) secured Jewish sovereignty in part of the former British mandate Palestine. The end of hostilities also marked the beginning of a large-scale national memorialization project of the war that involved bereaved families, local communities and state agencies (Sivan 1991; Azaryahu 1995). Poems mourned the dead and honored the heroic sacrifice of young men and women. A national day of remembrance for the fallen was instituted in 1951. One of the most popular vehicles for commemorating heroic sacrifice and victory in battle was the war memorial. Notwithstanding the particular interests and concerns of those in charge of promoting and building them, these memorials were about inscribing the story of the war into the landscape and they formulate the meaning of death in battle in terms of a patriotic legacy of heroic sacrifice (Azaryahu 1992; Shamir 1996).

The commemoration of Israel's War of Independence also involved the transformation of battlefields into battlescapes by means of memorials designed to evoke the memories of the battle in the place where it took place. Based on the notion of history as being an intrinsic quality of the local landscape, commemorative measures purportedly only make explicit that which is implicit in the local landscape. Such commemorative measures employ the agency of display to create an interpretive interface that mediates and thereby transform the landscape into a vision of history. As David Lowenthal observed, 'what is potentially visible is omnipresent' (1985: 239), which explains the power of landscapes to render historical memories of dramatic events that took place there tangible and concrete.

Monuments reify the past in the landscape (Foote and Azaryahu 2007). Their social relevance and vitality is correlated with their capacity 'to coalesce communal memories and aspirations and becomes a mechanism for the projection of personal values and desires' (Nelson and Olin 2003: 6). Of much public resonance were battle remains that, as tangible history, became venerated relics. Prominent among these were the wrecks of armored vehicles left alongside the mountainous road linking Tel Aviv with Jerusalem. These wrecked vehicles belonged to the convoys that defied the Arab siege of Jewish Jerusalem in spring 1948. In the patriotic lore of Israel's War of Independence, the battle for the road to Jerusalem figured prominently. Most notably, a postage stamp issued in 1950 featured Jerusalem at the apex of the road leading to it through ravines and hills.

Converted in patriotic culture into sacred relics, the wrecked vehicles epitomized the heroic sacrifice of the men and women who volunteered to join the convoys to the besieged Jewish Jerusalem. In their capacity as venerated relics, the battle remains at Sha'ar HaGai enjoyed high testamentary value: they served as both witnesses to and physical evidence of history. Moreover, for those aware of their story, the notion that they were left *in situ* projected a sense of time frozen at the end of battle. In their semiotic capacity as an interface between the historical drama and the contemporary landscape, the wrecked vehicles were not only conducive to transmitting commemorative meaning, but also to bridge the gap between 'now' and 'then'. As both witnesses and physical evidence, they belonged to the past they signify and their 'testimony' was rendered both credible and persuasive.

Perceived in patriotic culture as genuine remains of battle left in their original place, the relics exuded authenticity that substantially augmented their commemorative function (Azaryahu 1993). Recognized as belonging to the material fabric of history, the relics were instrumental in transforming the local landscape of the ravine where the road was built into an affective battlescape that conflated history and its commemoration in the local landscape.

Based on archival material and newspaper reports, this chapter explores the transformation of the wrecked vehicles along the road at Sha'ar HaGai into venerated relics and constituent elements of a distinguished battlescape of Israel's War of Independence. An underlying premise of the analysis is that the commemorative history of the relics involves three successive periods: preserving the past (1948–1955); staging the past (1956–1969); and rearranging the past (1970 to the present).

Preserving the past

In June 1948, during a month-long truce, the American journalist Robert St John traveled to Jerusalem in a car adorned with a United Nations flag. The road to Jerusalem was closed to civilian cars and the journey to Jerusalem was still considered unsafe. The description of the road is a first-hand impression of the situation prevailing there shortly after the battle for the road:

> What we saw in the ditches beside the road in the gully called Bab el Wad [Figure 5.1] told, better than history books will ever be able to tell, the price the Jews of Tel Aviv paid to get supplies to their brothers in the Holy City. There wasn't a hundred-foot stretch the whole way without its piece of wreckage. Some of the trucks which had hit land mines or been shot off the road into the ditches and set afire had been big ones. They lay here now like dead animals with their feet in the sky and their split-open entrails exposed. The wind that blew through Bab el Wad that day was strong enough to keep the wheels of one of the over-turned trucks constantly spinning, like the blades of a windmill.
>
> (St John 1949: 70)

Figure 5.1 Bab el Wad, 1948.

Source: Courtesy IDF and Defense Establishment Archives.

The surreal atmosphere that the author conveyed combined the human drama that had taken place and the almost grotesque sight of the remains of burnt-out vehicles abandoned alongside the road. The silence resonated with a sense of the extraordinary events. The only movement seen was the purposeless rotation of the wheels of an overturned truck. The reference to dead animals likened the local landscape to an abandoned slaughterhouse.

The uncanny sight of the wrecks was also featured in the Hebrew press as early as July 1948:

> You will see many of them when you travel today to Jerusalem. A long, double column planted alongside the road . . . those which did not arrive. Up-turned skeletons of cars, crashed, burnt, wounded in their heads and hips, having died from their wounds. At the twenty-first, second and third kilometer . . . a long row of monuments, bent, twisted wrecks of iron, frozen in spasms of agony. Grotesque ghosts cast their shadow on the road in the light of a pale moon.
>
> (Karlebach 1948: 6)

To avoid blocking the road, the wrecks were simply pushed to the ditches alongside the narrow, two-lane road at the bottom of the ravine and, as later explained, 'no one thought to move them away' (Shamir 1963: 3). The remains of vehicles left on the roadside had already been canonized in 1949 as emblems of heroic sacrifice: 'Each red-rusty skeleton – memory of life cut short; memory to human beings, most of them very young, who were burnt alive, or crushed or relieved from agony by an enemy's bullet' (Yevin 1949: 3).

'Bab el Wad', Haim Guri's song that combined the place, the material remains and remembrance of unnamed heroes was written in 1949 and immediately became a canonical text of Israel's culture of remembrance. In the song, Guri referred to the convoys, to the dead lying on the roadside and to the 'iron skeletons' (the armored cars) that were 'silent' like his dead comrades (see Figure 5.2 later in chapter).

According to an observation made as early as November 1949, the already popular song reflected 'the past, the near, painful, bloody and glorious past' (Keisari 1949: 3). In charge of the Hebraicization of the national landscape, in 1950 the Names Commission of the Prime Minister's Office assigned the name Sha'ar HaGai as the Hebrew name of Bab el Wad (the Hebrew designation is an exact translation of the Arabic name) (Azaryahu and Golan 2001). Yet thanks to Guri's influential song, Bab el Wad also persisted as a reference to the historical place associated with the heroic drama of resolve and sacrifice. Invoking the 'road to the city' provided a wider context of the battle for Jerusalem. The direct appeal to the place to 'Forever remember our names' invested the terrain with mythical meaning and transcendental purpose. The line, 'the iron skeleton is silent as my comrade' created a metaphoric linkage between the relics and the dead. The wrecks were silent, but as relics they had a powerful story to tell. According to cliché, the relics were a 'mute testimony' to the heroic past (Eichner 1996: 5). Yet on a deeper level of mythical interpretation, they were suffused with a permanent cry captured in their metal fabric: 'the wrecks, relics of the flames and the bullets, which are up to this day lying by the roadside . . . burnt skeletons whose quintessence is a cry – the cry of the open road' (Anonymous 1955: 17).

The evocative power of the relics in popular patriotic culture was manifest in references to them in the Hebrew press. In 1951 common knowledge had it that 'the skeletons of the cars at Sha'ar HaGai are the most sacred monuments to (Israel's) War of Independence' (Anonymous 1951: 2). Dubbed 'wrecks of glory' and exalted as 'artifacts of sublime and magnificent splendor', the wrecks were invested with sacredness (Dayan 1953: 2; Keisari 1958: 2). In retrospect, for those who took part in the battle for the road or were aware of the commemorative significance of the wrecks, the need to preserve them was 'a natural, spontaneous act, anchored in a stern will to leave some tangible testimony to those fateful days' (Shamir 1963: 3).

Raising the issue of whether and how they should be instituted as officially recognized memorials was the result of the already prominent role they played in the emergent landscape of memory of Israel's War of Independence. The idea of transforming the burnt-out and deformed wrecks lying along the road into official memorials to the heroic battle for the road to Jerusalem was first written down by the Ministry of Defense spokesperson in February 1950 in a letter he sent to the deputy Minister of Defense. In the letter he mentioned that the idea was circulating to place one of the wrecks on a concrete pedestal in order 'to preserve it as a memorial' (Letter 1950a). In his opinion, the idea was worth examining. Ostensibly unrelated, a few days later a short piece in *The Palestine Post*, Israel's only English-language newspaper and the only paper published in Jerusalem (it was soon renamed *The Jerusalem Post*), cited an unnamed reader's suggestion that 'the chain of wrecked and rusting armored cars and trucks lining the side of the Tel Aviv–Jerusalem road be cared for as a permanent monument to the heroes of the convoys' (*The Palestine Post* 1950: 2).

Albeit short, the report in *The Palestine Post* was the first to publicly address the issue of the fate of the burnt-out skeletons of armored cars dispersed along the road at Sha'ar HaGai. The publication had immediate consequences. The first was the notion that the lack of official caring for the relics indicated negligence and by implication, disregard for the heroic sacrifice of those who had died there. The second was the direct involvement of the Ministry of Defense in dealing with the fate of the wrecks and their possible performance as commemorations in the local landscape.

In a parliamentary question to David Ben-Gurion, Israel's Prime Minister and Minister of Defense, a member of the left-wing Mapam party addressed the need to care for the relics. In pathos-filled language he stated their commemorative meaning: 'Each skeleton of a car is tied to the name of warriors who fell in battle and sacrificed their lives in the war to liberate the homeland' (Protocol 1950: 1474). After stating the symbolic meaning and significance of the relics, he suggested that the alleged deplorable condition of the relics was 'a desecration of the honor of the fallen soldiers and a blunt obliteration of the memory of chapters of

heroism in the war'. He further proposed exhibiting the relics as official commemorations 'by building a fence around the sites where the wrecks are located,
to mark the sites with the names of the fallen and with a short description of their
endeavors'.

The short piece in *The Palestine Post* prompted leading figures in the Israeli
army (Israel Defense Forces – IDF) and the Ministry of Defense to consider the
issue of the relics and their fate. The Chief of Staff, Yigael Yadin, expressed his
support for the idea to convert the wrecks into official monuments (Letter 1950b).
However, he was against the idea of building a museum for the relics; in his view,
the wrecks should become monuments 'in the same place in which they are'
(Letter 1950c). Probably unaware of this position, the head of Israel's military
archive informed the Director-General of the Ministry of Defense about the
intention to transfer the remains of the vehicles to a 'war museum' and in addition
to build a memorial at Sha'ar HaGai from the remains (Letter 1950d).

Ben-Gurion's view in this matter was in stark contrast to Yadin's. In his response
in the Knesset to the aforementioned parliamentary question, he categorically
rejected the suggestion of transforming the wrecks into memorials:

> I do not agree with the whole of his proposal. The skeletons of the vehicles
> should be removed, and in the appropriate places memorial columns should
> be built. The commemoration of the war by way of proper arrangements
> throughout the entire country and in the approaches to Jerusalem – is still
> ahead of us.
>
> (Protocol 1950: 1474)

Ben-Gurion did not specify to where the wrecks should be removed. His prefe-
rence for monumental commemoration of the battle for the road to Jerusalem
resonated with ideas already developed by the Unit for the Commemoration of the
Fallen Soldiers (UCFS) in the Ministry of Defense. Yosef Dekel, the head of
the UCFS, recommended in 1950 building two national monuments to the war.
One was a monument to the Unknown Soldier in the projected government district
in Jerusalem (which, incidentally, has never been built). The other was a monument to those who fought in the battle for the road to Jerusalem, to be erected
along the roadside. The building of such a monument was approved in 1951 by
the Public Council for the Commemoration of the Fallen that, in its advisory capacity and in cooperation with the UCFS, was in charge of shaping the character and
form of national commemoration of the fallen in Israel's War of Independence.

From Dekel's perspective, and in accordance with Ben-Gurion's expressed
wish, the emphasis was on building a national monument to the heroes of the battle
for the road. However, at this stage of their commemorative history, the popular
appeal of the relics was a decisive factor in their preservation. In summer 1952, the
need to save the wrecks became acute, when it transpired that the Financial
Department in the Ministry of Defense was about to announce the sale of the
wrecks as defunct military equipment. For the official concerned, the wrecks had
value merely as scrap iron:

> We saw the wrecked vehicles along the road to Jerusalem as usable material, the same way we considered the parts of airplanes scattered all over the country and the military equipment damaged in the war. . . . [F]rom a sentimental perspective all these wrecks have the same background and the same symbolic value.
>
> (Letter 1952a)

The Ministry spokesperson was well aware that selling off the battle remains would provoke an outburst of anger (Letter 1952b). For Dekel, the building of the national monument was a priority. But he was also well aware that the fate of the relics was also an important issue. Dekel informed his superiors in the Ministry of Defense that 'bereaved parents and people of taste' told him that it was impossible to move the relics, and moreover, they should be preserved (Letter 1953b). Prominent among these 'people of taste' was the architect Yochanan Ratner, still serving as a general in the IDF. Ratner was enthusiastic about the commemorative possibilities offered by the wrecks:

> It is a shame to avoid an opportunity to create something extraordinary that will remind of something extraordinary, and to articulate it in manner that will convince future generations. . . . This situation has been created on the road to Jerusalem. It is difficult to conceive of something more dramatic. Wrecks of cars spread out along the road . . . lying until this day as soldiers injured in battle . . . after every curve there appears another 'monument', in strange forms that let us feel what happened to soldiers and drivers . . . there is nothing like this in the entire world. I do not suggest doing anything special in each of these 20 sites. On the contrary, less is better, and in the same form as they remained on the roadside.
>
> (Letter 1953a)

Ratner's advice focused on the unique possibilities offered by exploiting the symbolic capital of the authenticity exuded by the relics left on the roadside. He did not suggest any possible linkage between the relics and the official monument. However, participants in the public competition for the design of the national monument in Sha'ar HaGai published in January 1953 were informed that such an option existed. According to the guidelines to participants:

> On the roadside are wrecked vehicles . . . These are not only remains of vehicles, but markers to heroism that together with their environment are shrouded with sacred events ever since (these events) and for evermore. If possible, it is desired that the designer will also include a solution to these metal skeletons in his design.
>
> (Guidelines 1953: 5)

The result of the public competition for the design of a monument was announced in summer 1953 with no competitor winning first prize. The decision was to proceed with the design that won second prize: a sculptured figure, 12 metres high, carved in the rock, of a young man moving forward holding in one hand a hand grenade and in the other supporting his wounded comrade leaning besides him. This design avoided any reference to the relics. Critics claimed that the design was banal, a cliché that failed to capture the distinction of the local story of Sha'ar HaGai (Meisels 1953: 5). This design subsequently was never built in the local landscape.

In September 1954 Dekel informed the Defense Minister that no proposal had yet been made as to how to solve the problem of the wrecks by the side of the road to Jerusalem (Letter 1954). He explained that the problem was that while these relics had symbolic value only if left in their current place, they were consumed by rust and could not be preserved. He suggested a compromise solution: the relics should not be removed until the national monument planned by the Ministry of Defense was erected. The fact that the relics were already a well-known aspect of the local battlescape at Sha'ar HaGai was evident in summer 1955, when Shimon Peres, the Director-General of the Ministry of Defense, made a promise to the sub-committee for security and foreign affairs of the Knesset 'to build a monument at Sha'ar HaGai and to preserve the wrecked cars' (Letter 1955).

In stark contrast to the emphatic disregard for the relics Ben-Gurion expressed in 1950, the promise made by Shimon Peres, his protégé at the Ministry of Defense, put the monument and the relics on a commemorative par. However, in contrast to the impression created by the instructions of the Director-General of the Ministry of Defense, the UCFS role in shaping the local battlescape was limited to the monument only. In July and October 1956 two Knesset members each complained about the desperate condition of the wrecks alongside the road to Jerusalem (Protocol 1957: 1737). Evidently the relics belonged to the patriotic consensus and transcended party politics. A member of Herut, a right-wing party, suggested to collect the relics and to build a memorial. A member of Mapam, a left-wing party, reiterated the suggestion he had already made in the Knesset in 1950 to arrange the remains of the armored cars as a memorial. Ben-Gurion's response in the Knesset was that the Department of Landscape Improvement (DLI) in the Prime Minister's Office was in charge of 'positioning the wrecks alongside the road to accord with their importance'.

Staging the past

The DLI was established in October 1955 with the objective of developing tourist infrastructure by improving historical and archeological sites as well as sites associated with the heritage of Israel's War of Independence. Among the DLI's first projects was placing explanatory signs at various historical sites throughout the country, including Sha'ar HaGai (Anonymous 1956: 4). Based on and in accordance with the prevalent notions about the unique character of the relics, ostensibly the task of the DLI was simple enough: 'to leave the remains of the cars in their place and to protect them' (Letter 1957c). In actuality 'improving the landscape' in connection with the relics at Sha'ar HaGai involved applying measures to enhance their visual capacity to display and thus to relate the heroic past. However, tampering with the relics entailed undermining their rhetorical capacity to perform as authentic remains and hence primary witnesses to the heroic past.

The formation of an 'improved' battlescape was entrusted to two leading Israeli landscape architects, Lipa Ya'alom and Dan Zur, who were also engaged in other big projects that the DLI promoted. For the architects involved, 'improving the landscape' of Sha'ar HaGai meant staging the past: the underlying concept was of relics as artifacts displayed in the local landscape. This entailed three inter-connected issues. One was replacing local wrecks by others brought there for the purpose of their exposition as remains of war. The other was moving relics from the ditches where they were lying to a higher elevation along the road to make them more prominent in the local landscape and hence more visible to drivers on their way to and from Jerusalem. The third was to paint the artifacts to prevent corrosion.

Staging the past entailed its preservation. Already in 1952 it had been argued that with no preventive treatment, corrosion was bound to bring about the disinte-gration and gradual disappearance of the wrecks (Letter 1952a). Preservation meant intervention in order to halt the corrosive effect of time on the relics. The paradox was that in order to preserve the objects in their original condition they had to be constantly maintained.

When the official treatment of the relics as constituent elements of an official battlescape began in 1956 it became clear that some wrecks had already corroded to the extent that they could not be saved. A decision of much significance was to exchange those original relics for other wrecks made to look like battle remains. The possibility of replacing original relics with defunct armored cars that did not belong to the convoys was based on the notion that in principle it did not matter whether the physical remains were original or only appeared as such. Perceived as generic battle remains and without a documented history of their own, the original relics were rendered exchangeable by those in charge of improving the local landscape of heroic sacrifice.

In October 1956 the DLI approached the IDF with the request for a dozen defunct armored cars, which, in order to make them appear like authentic battle remains, 'should be burnt, pierced by shooting and be painted according to instructions' (Letter 1956). This request was approved, and the defunct armored vehicles stored at a military base that were chosen for their new career at Sha'ar HaGai were treated as requested in January 1957 (Letter 1957a, 1957b).

The preparation of substitute battle remains announced the first phase in staging the past at Sha'ar HaGai. In April 1957 a group of four renowned architects commissioned by the DLI visited Sha'ar HaGai to inspect the situation on the ground, and subsequently made detailed recommendations regarding the design of the local battlescape (Report 1957). The architects recommended painting the wrecks in dark grey. Importantly the relics were not rearranged in new sites but largely remained where they were along the road: relocations were limited to moving relics to a higher elevation or placing them with trees in the background.

Inspecting the actual condition of the battle remains, the architects decided that four original remains of armored cars had to be removed, while four 'new' wrecks were placed in four different locations. The total number of the relics on display remained 14, arranged in 8 locations alongside the road.

Removing original relics meant that they were stripped of their sacred aura and reverted to the status of indistinctive wrecks. Their positioning alongside the road ostensibly rendered the new armored vehicles emblematic of a heroic history they were not part of and allegedly invested them with testamentary power to events in which they did not participate. Yet the effect of such tampering with the relics was dependent on public reaction to the alleged 'inconsiderate treatment of these sacred relics' (Letter to the editor 1958b: 2). Notably, the public reaction was muted. Several letters to the editor that appeared in 1958 in conjunction with the celebration of Israel's tenth anniversary condemned the changes made in the local battlescape as 'Hollywood in Sha'ar HaGai' or 'a plastic surgery to the relics of Sha'ar HaGai' (Letter to the editor 1958a: 2, 1958c: 2). In particular, the alleged fact that the relics on display were not original but brought there from an army depot amounted to 'faking history for the sake of impression' (Letter to the editor 1958a: 2). Painting them was also criticized.

In response to the letters to the editors, the DLI explained that painting the wrecks was necessary to thwart corrosion; the vehicles were moved to a higher place since it was impossible to leave them near the road or in the ditches alongside the road, where they could hardly be seen (Letter 1958). According to the official statement, all remains of armored cars on display were deployed at Sha'ar HaGai in 1948. Asserting that there was no intention 'to change or correct history' (Letter to the editor 1958d: 2), the closest to disclosure was the claim that two armored vehicles were moved in 1948 to a military camp, but were 'returned' to Sha'ar HaGai by the DLI and that some wrecked civilian trucks had to be removed because of their rusty condition. Seemingly, the official explanations were convincing, and no new claims about 'faking history' at Sha'ar HaGai appeared in the press.

The conversion of the local scene into an official battlescape also entailed placing historical signs in conjunction with the relics. The DLI consulted with commanders who led the convoys in 1948 about where and when battles were waged along the road (Letter 1959). The dates of the battles were engraved on local rocks, which were placed where these battles had taken place. Engraving the date converted each rock into a primary witness, the assumption being that the rocks had remained *in situ* since the battle they 'witnessed'. The message encapsulated in the engraved dates was both laconic and powerful. Each inscription evoked a battle of which the heroic sacrifice of the relics was emblematic. Concurrently the dates made it clear that the wrecked cars were not mere scrap iron forgotten along the highway, but meaningful elements of the story of the battle for Sha'ar HaGai. Later on, standard brown-red markers with concise historical information were placed near the vehicles, signifying to passers-by acquainted with the role of such markers in the Israeli landscape that the armored vehicles lying there were of some historical importance (Figure 5.2). However, reading the information inscribed in the markers entailed visiting the site.

Figure 5.2 Markers: boulder with dates; a sign with Guri's song in Hebrew.

Source: Author.

Whereas the DLI was in charge of improving the local landscape, the Ministry of Defense promoted the building of the official monument. In 1961 a second public competition to design a monument was announced and, in March 1962, the winning design among the 57 submitted proposals was announced. The monument, in the form of a bundle of stainless steel pipes pointing toward Jerusalem, was inaugurated on March 15, 1967. The structure provided what the relics could not: the visual representation of the meaning of the sacrifice in abstract, albeit in easily comprehensible terms. The monument was constructed on a hilltop overlooking the road. The idea of removing the wrecks once the official monument was built had now become unfeasible as these wrecks had unequivocally become emblematic of the battlescape.

Nonetheless, the construction of the official monument modified the discursive context. The stainless steel structure did not purport to exude authenticity, a domain left solely to the relics. Yet, at the same time, the very lack of authenticity and the use of stainless steel implied the permanence of the monument in contrast to the relics, which needed constant care to prevent their disappearance from the scene.

Rearranging the past

The notion of the relics being left by the roadside at the end of battle was an essential part in their power of evoking a sense of time that had frozen in 1948.

The notion of 'frozen time' ignored the changes in the location of the wrecks. Obviously the wrecked vehicles had had to be moved to the roadside when the road was opened for regular traffic in summer 1948 (Letter 1977). They were moved again in 1957/1958, when the DLI was in charge of designing the local battlescape at Sha'ar HaGai. Later on, relocating the relics was a necessity whenever the road to Jerusalem was improved. Such relocations also prompted successive attempts to arrange the relics in the landscape in a manner befitting their mnemonic function and status as venerated relics.

The succession of relocations began in 1970, when massive development work was launched to convert the road to Jerusalem into a multi-lane highway. For the duration of the work the relics were moved away from their former locations alongside the road (Letter 1970). Criticism was directed against what looked like disregard for the relics. A report in a daily newspaper alleged that 'while reshuffling the hills and detonating boulders the relics were thrown sideward, scattered or piled with no apparent order, and no one cared to protest the insult' (A 1 a 1972: 7). A letter to the editor of the Israel Army magazine reiterated the idea that disregard for the relics undermined the memory of the heroes (Letter 1972b).

While the building of the new highway was underway, officials at the UCFS – again in charge of the relics – were engaged in devising a policy regarding the future arrangement of the relics alongside the road. An early idea was to relocate some relics 'and display them in more visible places above an ancient stone terrace' in order for them 'to be better seen' following the rerouting of a section of the road (Report 1970). Later on, the design of the local battlescape was entrusted to the landscape architects in charge of planning the post-construction landscape of Sha'ar HaGai. In principle, the stated goal was 'to preserve an authentic picture as far as possible' (Letter 1974). This meant that, as before, each relic was treated as a monument in its own right (Letter 1975). The main point was that, despite their relocation, their position alongside the road was maintained. Notably the architects favored concentration to dispersion. According to the new arrangement, the wrecked vehicles were concentrated in three groups only 'in a form that enables [people] to clearly see them while travelling uphill to Jerusalem or downhill to Tel Aviv' (Letter 1972a).

In 1983, following another widening of the road, the relics were again relocated (Letter 1983a, 1983b). In conjunction with construction work designed to improve the road connecting Tel Aviv and Jerusalem in February 1996, the 14 relics were transferred to a factory specializing in anti-corrosion treatment. They were brought back from renovation in April 1996, a few days before Remembrance Day for the fallen soldiers of the same year. However, the relics were not brought back to their former place. Underlying the 'rethinking' of the armored vehicles among their custodians was the principle that by using minimal financial resources it was possible to improve their location along the highway (Letter 1996). The result was a strategy of staging the past by means of enhanced authenticity: the relics were moved to five sites. Of these, three were in segments of the defunct old road, which had been exposed especially for the purpose of displaying the relics there. Authenticity in this case was a function of 'returning' the wrecks to the old,

Figure 5.3 A staged convoy.

Source: Author.

original road, though not necessarily to their original places, i.e. their location at the end of battle.

However, this arrangement was not the final resting place of the relics. In February 2009 they were again treated against corrosion. On this occasion a joint committee comprising the UCFS, the Jewish National Fund and the National Parks Authority made a strategic decision about how to rearrange the relics in the local landscape. The idea of the relics as being 'frozen in time' after they had been left by the roadside was abandoned in favor of their re-arrangement into two convoys, each comprising seven relics, along the road. Notably the arranged convoys were placed on concrete floors that simulated the road, while each vehicle was positioned on a small concrete pedestal. The armored vehicles were painted grey, and signs informed the history behind the arrangement (Figure 5.3).

The idea of putting each vehicle on a concrete slab with a sign relating the relic's story had already been raised in a letter to the UCFS in 1970 (Letter 1970). In 1972 the head of the UCFS explained why this design option had been rejected:

> We could however place the armored vehicles on concrete slabs and as a result we would have created a shiny museum artifact and nothing more. Yet the intention was to place the relics as they were seen after they had been stopped.
>
> (Letter 1972c)

In 1972 the underlying principle was that the positioning of the relics in the local landscape should sustain the impression that in connection with the relics, time had frozen in 1948, ostensibly suggesting that the wrecks had been left where they had been at the end of the battle. However, 40 years later the idea of placing the relics on a concrete 'stage' was realized in the local landscape. The difference was that in 2009 the relics had been arranged as convoys, which meant that they were disassociated from specific events and their location and made emblematic of the general story of convoys breaking their way through to the besieged city.

Concluding remarks

The fact that the wrecked vehicles left by the roadside were not removed from he local scene after the end of fighting in 1948 was decisive in their transformation into sacred relics. In 1950 the idea to transform the wrecked vehicles in Sha'ar HaGai into official monuments of a heroic chapter of Israel's War of Independence was for the first time discussed among government officials aware of the commemorative meaning, testamentary worth and popular appeal of the relics, which already in 1951 were characterized as '[t]he most sacred monuments to [Israel's] War of Independence' (Anonymous 1951: 2). In 2003 the pictorial image of a wrecked vehicle was chosen as the theme of that year's Remembrance Day postage stamp (Figure 5.4).

The power of the wrecked vehicles to induce remembrance was associated with and derivative of their perception as authentic relics; namely, that as genuine remains of battle left in their original place, the relics were 'tangible testimony' to the fateful events that caused their demise (Shamir 1963: 3). Yet the commemorative history of the relics following their institutionalization as constituent elements of the local battlescape is that of diminishing authenticity. In principle, beginning in 1956, staging the past involved a twofold relocation. One was relocation in time.

Figure 5.4 Memorial Day postage stamp, 2003: 'The armored cars on the road to Jerusalem'.

Source: Photo courtesy of the Israel Philatelic Service, Israel postal company. Graphic designer: Eva Cohen Saban.

This was the case with replacing corroded original relics with artifacts brought there for the purpose of display and made to look as if they were damaged in battle beyond repair. The other one was relocation in space, evident in the successive arrangements of the relics alongside the road.

Replacing some wrecks with others and the relocation of others alongside the road in 1957/1958 roused little public protest; surprisingly 'faking history', as one critic called it, did not challenge the authority of the relics as both 'witnesses' and 'evidence' to a heroic chapter of the War of Independence. As David Lowenthal notes: 'The removal of relics whose lineaments are indissolubly of their place annuls their testamentary worth and forfeits their myriad ties with place' (Lowenthal 1985: 287). Indeed later relocations of the relics made necessary by successive major works to widen the road undermined any sense that the relics had been left in their original place. Arranging them in two 'convoys' in 2009 was also a statement that the display of the relics alongside the road was a matter of landscape design and staged authenticity, where staged authenticity was not about faking history but about displaying its physical remains as a visual story of the past in the local landscape.

The symbolic capacity of the relics to relate the past is anchored in the cultural convention that accords them the role of both testimony and witnesses. The testamentary power assigned to the relic is evidenced in a piece written by an elementary school pupil in 1958 after visiting the place. Entitled 'What the armored vehicle told me at Sha'ar HaGai', this was not another conventional reiteration of heroic history, but a monologue of the 'rusty, wrecked armored vehicle':

> 'Oh,' cried the armored vehicle – 'where are you, my sons who sat inside me? How much I miss you! Look, my dear ones, now I am lying alongside the road, useless. But it was not like that when you were driving me. At that time I was among the conquerors of the road to Jerusalem. I recall the terrible day. The Arabs were shelling me. Yet my brave and heroic drivers were shooting back and hit the enemy. Oh, how many Arabs. Suddenly I felt a hard hit. I felt strong pain in all my organs. I felt heat and was shaken all over . . . and they, my drivers, were lying inside me and stuck their body to mine . . . later on they were taken away and we were separated . . .'.
>
> (Cited in Fleshner 1958)

The relic had been transformed into a witness and history into a testimony. As a witness, the relic was rendered human: it was sad, it sighed, felt pain and heat, and more importantly, it could 'speak' with its own 'voice' and 'tell' what it 'witnessed'. The monologue of the armored vehicle contrasted with the canonical text of Haim Guri's song, where the relics were silent as the dead and the demand to remember the names of the fallen was directed to the battlefield that is mentioned by its name: 'The metal skeleton is silent as my comrade/Bab el Wad!/Forever remember our names.' Notably, in both texts the dead were nameless.

Obviously, the power of the relics to 'testify' entails prior knowledge of the local history. Relics do not speak for themselves. Their ability to 'testify' is predicated on associating the relics with their (hi)story. The case of an American reporter who claimed in 1954 that the wrecks alongside the road to Jerusalem were defunct articles of American aid is illuminating; conspicuously, the understanding that the relics were susceptible to be confused with scrap iron prompted the urge to remove them (*Zmanin* 1954: 3). It was also argued that new immigrants or young people unacquainted with the history of Israel's War of Independence were also oblivious to the historical meaning of the relics. For some they were no more than the wrecks of cars abandoned by their owners after crashing them (Ben-Shahar 1956: 2). The concern about people being unaware of the historical significance of the relics found its expression in repeated requests to place explanatory signs near the relics. However, a major problem was that, as there was no access to the relics, it was impossible for drivers and passengers to decipher the signs and the message was limited to the dates of battles engraved in the white rocks.

Beyond prior knowledge about the association of the relics with history, their interpretation and evaluation in terms of memory and legacy is a matter of perspective based on particular ideological premises. Writing about his bus ride from Tel Aviv to Jerusalem in 1994, the travel writer Paul Theroux mentioned these relics: 'Old-fashioned armored cars and rusty trucks had been left by the roadside as memorials to the men who had died in what the Israelis call the War of Liberation. The vehicles, so old, so clumsy, roused pity' (Theroux 1994: 393). To Theroux the relics 'roused pity'. In Israeli patriotic culture they have been associated with heroic sacrifice, evoking veneration and respect. Despite their repeated relocations in the local landscape, the authenticity they exude substantially augments their symbolic capacity to conflate the historical battlefield at Bab el Wad and the contemporary battlescape at Sha'ar HaGai.

References

A-l-a. 1972. On the margins. *Davar*, March 14 (Hebrew).

Anonymous. 1951. From the voter's notebook. *Ma'ariv*, July 23 (Hebrew).

Anonymous. 1955. Memorial to Sha'ar HaGai. *Ba'makhane*, November 30 (Hebrew).

Anonymous. 1956. Hundreds of signs for historical sites. *HaZofeh*, August 27 (Hebrew).

Azaryahu, M. 1992. War memorials and the commemoration of the Israeli War of Independence 1948–1956. *Studies in Zionism* 13 (1): 57–77.

Azaryahu, M. 1993. From remains to relics: authentic monuments in the Israeli landscape. *History & Memory* 5 (2): 82–103.

Azaryahu, M. 1995. *State Cults. Celebrating Independence and Commemorating the Fallen in Israel 1948–1956*. Sedeh Boqer: Ben-Gurion University of the Negev Press/Ben-Gurion Research Center (Hebrew).

Azaryahu, M. and Golan, A. 2001. (Re)naming the landscape. The formation of the Hebrew map of Israel, 1949–1960, *Journal of Historical Geography* 27 (2): 178–195.

Ben-Shahar, M. 1956. Signs will tell. *Herut*, July 31 (Hebrew).

Dayan, S. 1953. Wrecks of glory. *Ma'ariv*, September 4 (Hebrew).

Eichner, I. 1996. The armored cars are overhauled. *Yediot Ahronot*, March 3 (Hebrew).

Fleshner, A. 1958. *Moladti: Bulletin of State Primary School A, Nes Ziona*. Archive of Jewish Education, Tel Aviv University, 11/ 3.155 (Hebrew).

Foote, K. and Azaryahu, M. 2007. Toward a geography of memory: geographical contribution to the study of commemoration and memory. *Journal of Political and Military Sociology* 35 (1): 125–144.

Guidelines to the Public Competition for the Building of the Monument in Sha'ar HaGai. 1953. The Sha'ar HaGai file, Archive of the UCFS (henceforth AUCFS).

Johnson, N. 1995. Cast in stone: monuments, geography and nationalism. *Environment and Planning D: Society and Space* 13: 51–65.

Karlebach, A. 1948. The British plan. *Ma'ariv*, July 5 (Hebrew).

Keisari, U. 1949. Twilight. *Ma'ariv*, November 7 (Hebrew).

Keisari, U. 1958. Glory and its place. *Ma'ariv*, June 24 (Hebrew).

Letter. 1950a. Y. Ziv-Av to S. Avigur, February 14, IDF Archive, 62-20/1960.

Letter. 1950b. M. Avizur to P. Sapir, February 26, IDF Archive, 211-220/1970.

Letter. 1950c. M. Avizur to P. Sapir, March 8, IDF Archive, 211-220/1970.

Letter. 1950d. P. Sapir to M. Avizur, March 5, IDF Archive, 211-220/1970.

Letter. 1952a. M. Orbach to Y. Dekel, July 22, IDF Archive, 213-230/1970.

Letter. 1952b. I. Harrusi to Y. Dekel, August 19, IDF Archive, 213-230/1970.

Letter. 1953a. Y. Ratner to Y. Dekel, January 23, IDF Archive, 156-90/1972.

Letter. 1953b. Y. Dekel to S. Peres, January 29, IDF Archive 156-90/1972.

Letter. 1954. Y. Dekel to P. Lavon. September 15, IDF Archive, 157/72/90.

Letter. 1955. S. Peres to Y. Dekel, August 24, IDF Archive, 81-849/1973.

Letter. 1956. Y. Yanai to the IDF, October 18, State Archive LG 4/3834.

Letter. 1957a. Y. Yanai to L. Ya'alom and D. Zur, January 17, State Archive LG 4/3834.

Letter. 1957b. IDF to Y. Yanai, January 18, State Archive LG 4/3834.

Letter. 1957c. D. Levinson to Shelev Transportation Cooperative, May 20, State Archive, GL 4/3834.

Letter. 1958. D. Levinson to *Al HaMishmar*, March 26, State Archive GL 4/3834.

Letter. 1959. D. Levinson to D. Namir, October 22, State Archive GL 4/3834.

Letter. 1970. U. Shertzer to R. Avner, September 13, AUCFS.

Letter. 1972a. S. Seri to the Public Relations Department, Ministry of Defense, March 17, AUCFS.

Letter. 1972b. S. Fleshner to the editor of *Ba'makhane*, August 3, AUCFS.

Letter. 1972c. S Seri to S. Fleshner, August 31, AUCFS.

Letter. 1974. S. Seri to H. Rubinstein, September 8, AUCFS.

Letter. 1975. Y. Schwerdlin to Y. Lesser, August 31, AUCFS.

Letter. 1977. Y. Schwerdlin to K. Grossman, January 23, AUCFS.

Letter. 1983a. A. Einsten to U. Navon, June 10, AUCFS.

Letter. 1983b. A. Einsten to U. Navon, July 12, AUCFS.

Letter. 1996. I. Bernstein to U. Navon, March 20, AUCFS.

Letter to the editor. 1958a. *Al HaMishmar*, March 12.

Letter to the editor. 1958b. *HaAretz*, June 17.

Letter to the editor. 1958c. *HaAaretz*, June 23.

Letter to the editor. 1958d. *HaAretz*, July 24.

Lowenthal, D. 1985. *The Past Is a Foreign Country*. Cambridge: Cambridge University Press.

Meisels, M. 1953. The competition for the memorial at Sha'ar HaGai. *Davar*, August 28 (Hebrew).

Nelson, R.S. and Olin, M. 2003. *Monuments and Memory, Made and Unmade*. Chicago and London: Chicago University Press.

Protocol. 1950. Session #153 of the first Knesset on May 24, *Protocols of the 1st Knesset*.

Protocol. 1957. Session #284 of the third Knesset on April 9. *Protocols of the 3rd Knesset*.

Report. 1957. April 23, State Archive, LG 4/3834.

Report. 1970. *Report on a Visit to the Site*. December 27, AUCFS.

St John, R. 1949. *Shalom Means Peace*. New York: Doubleday.

Shamir, A. 1963. How long? *LaMerhav*, June 21 (Hebrew).

Shamir, I. 1996. *Commemoration and Remembrance: Israel's Way of Molding its Collective Memory Patterns*. Tel Aviv: Am-Oved (Hebrew).

Sivan, I. 1991. *The 1948 Generation: Myth, Profile and Memory*. Tel Aviv: Ministry of Defense Publishing House.

Smith, A.D. 1991. *National Identity*. Harmondsworth: Penguin.

The Palestine Post. 1950. February 23.

Theroux, P. 1994. *The Pillars of Hercules. A Grand Tour of the Mediterranean*. Harmondsworth: Penguin.

Yevin, Y.H. 1949. On the road to Zion. *Herut*, March 18 (Hebrew).

Zmanin. 1954. September 10 (Hebrew).

Part II
Narrative memorial practices

Storytelling and materiality
in placing memory

6 Who were the enemies?

The spatial practices of belonging and exclusion in Second World War Italy

Sarah De Nardi

The Second World War in Italy: memory as a dwelling practice

The Italian Resistance during the Second World War was *the* iconic episode shaping political and intellectual debate in post-war Italy like little else before or since. How the Resistance is remembered today varies due to myriad factors concerning social, political and cultural identity constructions (Forlenza 2012). Far from being a unifying force as purported by left-wing party propaganda (Battaglia 1953), the Resistance represented instead a deeply divisive event that still engenders profounds fractures in Italy's popular consciousness (Pezzino 2005). As well as an opposition to German occupation, the Resistance also implied a civil war (Pavone 1991; Behan 2009) in which anti-Fascists, many of them Communists, fought the Fascists to eradicate Mussolini's regime and to re-establish democracy in Italy. This resistance was significant in military and social terms: by some estimates 300,000 Italians were involved in direct action against the Nazi–Fascist regime, with many more supporting the fighters with aid, provisions and assistance. Some calculate that in those frantic 20 months (September 1943–May 1945) as many as 70,000 fighters were casualties, with a further 40,000 wounded (Lewis 1985: 23). The complex, often contradictory manner in which people remember the events unravelling during those 20 months can illuminate the impact actions, persons, interactions and perceptions can have in any given historical period, and what their implications are likely to be for postmemory (after Hirsch 2008).

Understanding memories of the conflict as dwelling practices helps us contextualise the formation and rehearsal of identities and how these were understood and, at any rate, a clearer focus on the intersection of place and identity encapsulated in Second World War stories is useful. In my research with veterans (2009–2013) I explored their engagement with places and persons during the Resistance and occupation. I used original and secondary archival sources and oral history interviews (conducted by myself and others) to explore the ways in which identities were performed and how strongly actors were linked with a certain locale by their identity, and what role they had played within a certain community or landscape. I mined sources for traces of a 'sense of place', a feel for how memories and impressions become embedded in places. My use of the term here refers to the mechanisms and expedients through which we, as human beings, live in the world around us: I understand sense of place to be a product of the 'dwelling perspective' (Ingold 1993) where we make, unmake and live in landscapes in co-optive (Ingold 2000: 175) interpersonal and meaningful ways. For Ingold, the landscape is 'constituted as an enduring . . . testimony to . . . the lives and works of past generations who have dwelt within it, and in so doing, have left there something of themselves' (1993: 153). The emplacing of memories in a region, community or town, can therefore be understood as a fluid, continuous process taking place over time: it is a polyvocal and open-ended process of world-making. Intriguingly, my work with veterans and their families in Italy suggests that the war's most disturbing aspect was the *disruption* of long-established dwelling practices – a disruption given by nasty, unnatural events happening in one's own domestic arena to friends and acquaintances. Scary and often unsettling occurrences took place in villages, towns and neighbourhoods where acquaintances and relations eyed one another suspiciously, almost as if playing a chilling game of hide-and-seek. This disruption or corruption of dwelling practices embeds negative emotions in memory, which in turn contribute to new dwelling practices and so on. Specifically, then, this chapter contextualises the dwelling practices of Italians living through the civil war and occupation of 1943–1945, situating tropes of violence, terror and estrangement that were sometimes metabolised as aberrant events disrupting the status quo – agreed upon local dwelling practices – but most often rejected and feared as expression of an alien Other. Italian communities who had lived under Mussolini's Fascist regime for 20 years had hoped for a quick end to the war in Italy after the King and Prime Minister Badoglio signed the Armistice with the Allies on 8 September 1943. When Hitler sent a special taskforce to spring Mussolini from prison on 14 September 1943, the latter's return to power under the aegis of the Reich a week after the Armistice (Battaglia 1953: 34 ff.) led to a widespread sense of loss among communities who felt emotionally torn between resource-hungry Partisan formations and the Fascists' renewed demands for loyalty and manpower. Yet in the copious literature devoted to the Resistance and occupation there has been scant mention of that heightened emotional and mental state of either side in the conflict: this chapter seeks to redress the balance in this sense. Yet, Ingold reminds us,

> [t]o perceive the landscape is . . . to carry out an act of remembrance, and remembering is not so much a matter of calling up an internal image, stored in the mind, as of engaging perceptually with an environment . . . pregnant with the past.
>
> (1993: 152–153)

Although this chapter title asks *who* the enemies were the question might well extend to '*where* were the enemies?'. Two fundamental elements of a dwelling perspective, that is, identity and sense of place (and a sense of where one is in relation to the world) became dramatically interwoven in the Italian Resistance episode. An engagement with those events entails an encounter with a number of tangible and intangible elements – places, other people, traditions, sensations, impressions – that make us feel 'at home' or otherwise (Ingold 1993, 2000; see also MacLure 2008). Indeed, a recent body of scholarship is increasingly engaging with more-than-representational understandings of sense of place and identity. For Thrift, identity and belonging pulsate with affective energy (2004: 57). We might also infer that belonging, feeling at home or feeling unfamiliar (Askins and Pain 2011; Valentine and Sadgrove 2012), had an immense impact on those living through the war years in Italy, and beyond.

While in this chapter I draw on Ingold's dwelling perspective (1993, 2000), these analytics also encapsulate relational insights given by theories of infra-humanisation in social psychology. A focus on dwelling practices as the building blocks of conflict remembrance based on interpersonal interaction suggests that perception of the 'enemy' during the war depended on the place-specific factors of inclusion and exclusion, on belonging or being an outsider, as well as on political loyalties. These dichotomies (by no means mutually exclusive) tended to colour the way that the Resistance was experienced and remembered in the decades after the events. In this sense, theories of infrahumanisation can contribute a great deal of depth to a dwelling perspective of the conflict in Italy as they unpick the mechanisms by which we perceive the human potential or compatibility of fellow human beings and how (and where) we locate them accordingly. Leyens et al. (2000) and Haslam et al. (2005) postulate that if we do not deem another person capable of experiencing truly human emotions (what the authors identify as 'secondary emotions'), we tend to dehumanise them; this 'infra-humanisation' therefore makes it easier to ignore others' pain, their need for help, and, in some extreme circumstances, enables us to perpetrate violence against others, which we would not be able to stomach if the person in question belonged to our 'ingroup'. The implications of the infrahumanisation of others in a conflict context is extremely important; this interactional paradigm can reveal how some individuals perpetrated horrific acts of violence against some, while demonstrating kindness and compassion towards others. By acting according to perceptual schemes of inclusion and exclusion, individuals end up creating dwelling practices by which they either live alongside persons of their kind, or are confronted with the displacing otherness of outsiders – and react accordingly (Paladino et al. 2002).

Sense of place, intended as the way(s) in which we feel at home among others or exclude them, is a paramount factor in dwelling practices. This chapter, therefore, explores several instances in which belonging, exclusion and othering of individuals and groups lead us to a greater understanding of how a historical moment such as the Resistance is remembered today. The memories explored in this chapter constitute the dwelling practices of persons who experienced the conflict in Italy first-hand, and for that reason I decided to include verbatim quotes from diaries, memoirs and interviews with war veterans and their families.

The rationale of this collection focuses on a critical unpacking of representational and Cartesian views of places and events; this conceptual scrutiny, in turn, brings to the fore the complex nuances of lived experience, memory, identity and sense of place (see Ingold 2000; Ahmed 2004; Saunders 2004; Navaro-Yashin 2012; Wetherell 2012). An emphasis on affect and on materialities is indeed thriving in recent work on conflict anthropology and cultural geography. New thinking on the role of objects and places as 'melancholy' (Navaro-Yashin 2012), as fleshy ensembles (Saunders 2004, 2009) and as 'sites of feeling' (De Nardi 2014b) has led to an engagement with experiential and material-cultural experiences of recent war and conflict (see also Moshenska 2008; Henig 2012). The investing of places and locales with emotions and mental states associated with animate beings qualifies landscape as a primary actor in the experience of war and Resistance. For example, during my engagement with 28 Resistance protagonists (De Nardi 2014a, 2014b), I often encountered the dystopian attributes of 'eerie', 'terrified', 'petrified' and 'out of place' attached to the affective ingroup of freedom fighters (Partisans), to Fascist militias, to spies, to landscapes and topographical landmarks such as woodland or a path ('treacherous woodland', 'mean mountains'). This is hardly surprising considering that geographical and territorial knowledge were vital to the success of the Resistance as a movement, to the smooth running of a network of activities and acts of resistance, and that resistance was a process with strong local and emotional connotations. Most importantly for the scope of this chapter, a dwelling perspective shapes the conscious or unconscious decision about who belonged (or not) in some place or to a group (see also Haslam et al. 2005). When did one feel 'out of place'? This expression, which we often use unthinkingly, reflects both an emotional and physical positioning in the world (Ingold 2000: 45 ff.). This consideration brings us back to the case study at hand: wartime Italy. In an occupied warzone, identities are perforce blurred, but my research shows the extent to which sense of place and identities were formed. Political consciousness, collaboration and fear of reprisals were not alone in shaping senses of places or allegiances – the 'glue' holding these multifarious agents together in experience, as well as in memory, was constituted by dwelling practices. Such practices included aspects from the personal (alienation, loneliness, homesickness) to the collective (displacement, othering and dehumanisation) and determined who was 'out of place'; in its shaping of perception and action, a dwelling perspective was fundamental in the development of resistance movements, to the establishment of regimes and to how these are remembered, above and beyond the Italian or Second World War case studies offered here.

Ingroups and outgroups during the occupation and civil war of 1943–1945

A dystopian affective regime (see Reddy 2001) and a traumatised emotional community (after Rosenwein 2006) prevailed in the north and centre of the Italian peninsula after the Armistice of 8 September 1943. Fear of capture, anxiety for loved ones involved in underground Resistance activities or fighting, or relief when the enemy departed, shaped moods and atmospheres (see Ankersmit 1996; Nordstrom 1997). By considering sense of place and identity as interconnected dwelling practices we begin to see the tapestry of good versus bad loosening, coming apart, and stray threads opening up unexpected snags (see Jackson and Mazzei 2011: 34). Why? When considering the ways in which people negotiated their experiential worlds, we often find perplexing results. My case studies suggest that the traditional 'good guys' (the Partisans) could behave and be perceived as strange, hostile or disaffected outgroups, whereas the traditional villains (the Fascists) could feasibly possess a quality of warm, respectable and father-like 'ingroup' belonging. The reasons determining the perception of one or the other depended, I argue, on a certain community's dwelling perspective, and on what place the individual or group in question occupied therein: in or out.

An analytics focusing on dwelling as the core of identity practices may allow us to understand wartime events and attitudes in terms of overarching local traditions and perceptions. For example, long-held grudges against Austrians and German-speakers in the greater Veneto region (Figure 6.1) after the Great War (Vendramini 1984: 175) may go some way towards explaining why the populace hated the Germans even when they were still Axis allies, or why the majority of Italians mistrusted and rejected German-speaking soldiers even when they behaved fairly (Pavone 1991: 206). Actors in the events of the Resistance

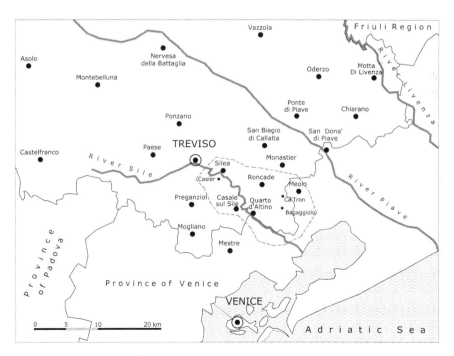

Figure 6.1 The Veneto region.

Source: Author.

positioned themselves in relation to others and to the places and landscapes in which they (and the 'Other') dwelled. Within the networks and social worlds of resistance and civil war, identities were shaped, adopted, rejected and feared. Although the mechanisms by which such identities were formed and rehearsed have never come under proper scrutiny, ignoring the protagonists' perceptions of 'self' and 'Other' oversimplifies conflict dynamics by flattening an otherwise complex lifeworld. More to the point, a political stance could not be taken by individuals without a corresponding construction of, and confrontation with, the 'enemy'. To be sure, the 'othering' of experience in memory could sometimes serve as a palliative relief from exposure to particularly awful events and memories (Radstone 2007; Cappelletto 2003). The dominant remembrance narratives about the Second World War contend that atrocities happened to one's lot because of external forces in a narrative that excludes the Other as the perpetrator (Cappelletto 2003). When violence and 'wrongness' were instigated by members of one's ingroup (after Leyens et al. 2000), their motivations were not questioned because they were members of said ingroup. Rather, aberrant elements *within* a group were 'othered' into oblivion, and ejected from one's affective network for being out of place. We might, then, attempt to probe into dysfunctional dwelling practices of actors implicated in the events of the Resistance in the hope of better understanding the conflict's disorienting cultural impact.

A Resistance fighter who had escaped from a prisoner of war camp after the Armistice, Liviano managed to join a Veneto Partisan brigade where he felt like an outsider among kindred spirits; he perceived himself to be out of place in a context that should have offered protection, fraternity and reassurance. Unfortunately for him, he did not fit into the local dwelling perspective and cultural schemata. Understandings and perception are negotiated in the encounter between different beings, and 'the outcome of each encounter depends upon what forms of composition these beings are able to enter into' (Thrift 2004: 62). Individuals upsetting established dwelling practices (even unconsciously and inadvertently) are constructed as aberrant elements even within closely-knit ingroups, which in this case resulted in a feeling of displacement, as Liviano recalls:

> I was a fugitive. The boys in the band offered me *puina*, which of course I had no idea what it was. I was starving so accepted it, and it turned out to be a wedge of delicious ricotta cheese . . . I decided to stay with them for a few days while I planned my homebound route. Apart from Renato, a welcoming, warm-hearted person, I am sorry to say that the others were stiff, obtuse, hostile, maybe due to a typical mountain-folk reserve. They gave me the silent treatment. I was *really unhappy* there, so much so that I regretted escaping from the Barracks, I longed for my superiors, particularly an Austrian Marshal who liked me and was kind to me.[1]

Liviano was made to feel like an unwelcome newcomer in the mountain Partisan formation he joined after his escape from a prison camp. His sense of alienation was most likely due to the fact that the men he encountered were north-ern mountain folk and he was a southerner – a cultural outsider unfamiliar with local dwelling practices. He literally did not *dwell* in those parts. But, aside from geographical differences, the two parties were unable to communicate well owing to Liviano's lack of understanding of the local dialect. Rejection and distancing mechanisms such as these stuck in the memory of the individuals who experi-enced them. The powerful makings of mistrust and instinctive rejection of others were not confined to geographical belonging, however. Elsewhere we learn that even local individuals were subjected to suspicion when they diverted in any way from the norm dictated by a specific dwelling perspective. Tito Spagnol, an older Partisan among young lads, felt that he was being watched with a mixture of fas-cination and concern about his suitability to a guerrilla lifestyle; probably, he later confesses, this suspicion arose 'because I was not like them' (Spagnol 2005: 41). Further, he wrote that 'they looked at my white hair, my walking stick, the liver spots on my hands' (2005: 38). He, too, was out of place (see Leyens 2009: 807). He was not part of the younger men's worldview, as if his age and diminished physical ability had erected an actual barrier between himself and the others. Tito's bodily otherness did not resonate with the group's expectations of how a Partisan should look, act and *dwell* in that specific landscape.

Spies, Fascists and the Germans: not quite 'right'

During the Italian war, identities were rehearsed, made and rejected according to dwelling practices perforce embedded in place. Place is not only a palimpsest of identities and practices but can also possess a counter-current of destructive energy: places can be unmade too. Destruction, destabilisation and alienation lead to the unmaking of reality and the tearing asunder of communities' dwelling perspectives. Below, a ritual Fascist practice recounted in an interview by Pavan illustrates an attempt to destroy, to destabilise and to unmake place:

> They [the Fascists] arrived on a jeep. They . . . went into a restaurant during a party, grabbed the female guests and shaved their heads.
> Q: Why did they go into the inn?
> Eh, they never liked it when people partied and had fun.
> Q: Were they not rather looking for Partisans . . .?
> No, they weren't looking for Partisans.
> Q: How long were they in the restaurant?
> An hour or so, then they left.
> Q: Did they eat and drink?
> No. they just shaved heads.[2]

The last fragment reveals another facet of the civil war in Italy: the misuse of places for violent ends. An inn, a place traditionally associated with pleasurable activities, was chosen by the Fascists as the site of a petty hate crime against civilians. The dwelling practices of that place were rewritten through violence and, by rearranging the unwritten perceptual coordinates of a familiar site, the Fascists cast themselves firmly out of place.

As well as a heavy military or paramilitary presence in the landscape, spies were rife and moles often infiltrated the Partisan rank and file. By providing stolen intelligence to the German and Fascist authorities for financial gain, spies effectively disrupted the established sense of place by creating a treacherous geography of betrayal and fear among the civilian population and the armed Resistance. In his memoirs, Tito Spagnol (2005: 24–26) recounts the story of one such person:

> The Platzkommandatur's[3] interpreter Virdis pretended to be half German stating that he had a German mother; only, we found out that he was all-Italian. Virdis was officially an interpreter but also acted as a ruthless Gestapo agent. He was a very attractive chap and had a way with women. I am pretty sure he seduced a few ladies just so he could gain information that would lead to our capture and execution.

In the above excerpt the infrahumanisation of an Italian who sells his services to the invading Hun (as the Germans were known among the Partisans) was clear-cut. In his decision to betray his fellow Italians, the man's noxious otherness made him almost impossible to relate to: an Italian, pretending to be German in order to ingratiate himself with the higher Nazi echelons! His very existence only made sense to Spagnol (the writer) when framed in an imposter narrative (pretending to be half German, pretending not to be a Gestapo agent, making love to women with ulterior motives). Virdis' construction as the destabilising Other is based on dissimulation and make-believe; as such, this person's identity as Italian, or his status as ingroup member (see Leyens et al. 2000: 190–191) does not have to be questioned because he does not live, think or feel the same way as Tito and his companions. This dualism places the interpreter out of place – neither here nor there (see Leyens 2009: 808) and reminds us how eliminating spies was relatively unproblematic even when they belonged to one's ingroup (see Brescacin 2014).

Overall, Germans and Fascists incarnated all that was bestial and inhuman about the war (Pavone 1991). They were perceived as being out of place or, even, outside of the natural order of things: they acted in ways in which no decent human being would accept; they set fire to homesteads, farmhouses, or raided little shops; they tore down barns, or hanged loved ones from millennial trees in the middle of a hamlet's square. From the Fascist perspective, on the one hand, anti-Fascists and their helpers were not the same, and therefore liable for violence in view of their otherness. On the other hand, violent acts on the part of the Fascists are recounted by veteran fighters and activists without a real attempt at explanation or analysis, as if the acts had been simply the doing of abject and not-quite-human beings:

> They had fun aiming at people. They killed two or three people.
> Q: What happened? Who was it?

The Fascists, the *Brigate Nere*, were blind drunk, and were riding around in a lorry. They had fun, driving around in this lorry; they sat with their rifles sticking out of the windows and they saw a little kid in the middle of a field. They took aim and shot him. They aimed at anything that moved.[4]

Another civilian called Anna recalls:

We were terrified of the almost daily searches. We saw them [the Germans] coming up from the road and then split up in the woods. They were after Partisans. And reprisals, oh god, those were the scariest. They killed my uncle during a reprisal. Little did they care about his children or his wife; they left his body out there like that of a stray dog on the ground, crawling with flies for two whole days. Calling German soldiers 'animals' would be unkind to the animal kingdom.[5]

The perceived infrahuman displacement of Germans also took another, more uncanny form. I recall an extraordinary story where a Partisan called Giorgio recounted a German retreat in the eerie tone and storytelling style sometimes reserved for an unexpected encounter with a strange beast:

Sometimes the Germans had an officer who wasn't the usual rotten lowlife and they didn't push too hard [against us Partisans]. Once or twice they came so close that they could have slaughtered us, but didn't. Once we were on a hilltop, and the special woodland SS squad found us. They blended in with the snow in their white uniforms and skis. At any rate, one of us was outside our hut, shaving by the stream. Wait for this! . . . When I joined my companion I saw that some twenty metres away, behind a tree, a white-clad German kept peering out as if unsure about what to do. He could have shot at us, burnt down our hut and run, but didn't. He saw us and quietly fled. They just . . . went away.[6]

Another similar story from an interview with Lorenzo goes as follows:

One time the Fascists and Germans were in a very remote mountain area, isolated, they were terrified of us Partisans, well, they always kept their lights on at night, flashlights and so on. Once the Allies mistook those lights for our drop-site signals and launched a whole load of supplies right in the enemy's lap.[7]

We might infer from the above passages that, despite the chasm between the dwelling practices of the two warring sides, the enemy – a puzzling unknown quantity – was still construed as capable of feeling afraid (see Leyens 2009), but in the sense of a primal fear that people acknowledge in both animals and humans (Leyens et al. 2000: 187). In engaging with others' fear, however, individuals also distance the enemy from normality. Cappelletto has aptly argued that most people do not conceive of the emotions of fear and danger as being part of the regular lifecycle (2003); she also suggested that violence carries an alien quality, which is instinctively explained through the placing of blame on outsiders. The fact that the Germans and Fascists behaved in bestial ways, positioning themselves out of familiar ingroups and challenging familiar dwelling practices, rendered them infrahuman: not quite 'right'.

A rock and a hard place: the outsider-Partisan and the insider-Fascist in Roncade

Some communities, however, clustered around an unlikely Fascist 'leader' in order to survive rather than out of ideological or political conviction. The history of the Resistance in the Roncade and Sile provinces of the Veneto region is politically and socially anomalous, and its messy ambiguity aptly illustrates our discussion of infrahumanisation in the context of local dwelling practices. When contextualised amongst the more traditionally 'polarised' and politicised social dynamics of most northern towns and countryside in the Veneto piedmont (see Figure 6.1),[8] we find here a pronounced blurring of the line between Resisters, active or passive, and Fascists and their collaborators. In Roncade, as we shall see below, the two 'categories' lived side by side almost seamlessly for one very simple reason: the local community relied on the industrial activities and patronage of the Menon family business (an automotive plant) for its stability – and the Menon family were sworn Fascists.

The head of the Menon family, Guglielmo, is presented in the book mentioned below as an ambiguous figure of both fatherly and sinister qualities. He was loved as much as leaned on financially in the close-knit local community of less than one thousand souls; a place in which he and his family filled an active social role in traditional dwelling practices based on trust. Tragedy struck when members of an outsider-Partisan brigade assassinated Guglielmo and his right-hand man, his brother Carlo – allegedly to set an example for the benefit of the local population. Oral historian and journalist Favero, and the many voices relating the events in his 2003 book *Inesorabile Piombo Nemico*, offers an insight into the ambiguity and multiplicity of local history and the vagaries of recollection. The Italian title (which translates as *Unstoppable Enemy Fire*) plays on the ambiguity of the wording on young Partisan Ugo Rosolen's memorial tombstone, as it is still unclear how he came to be captured and executed by Fascists in a town where he had never set foot before. The 'enemy' in the inscription is open-ended, and does not refer explicitly to insiders, outsiders, spies or enemies.

In Favero's book, the truth about Rosolen and the Menon murders is never fully explained; the reader is instead presented with fragments of explanation as to how exactly the Menons' assassination took place, who the parties responsible were, and how the population reacted to it. In some versions the murder led to the locals' hostility towards the Resistance because of one single event that threw a whole community into chaos. In other versions the populace accepted the fact as an inevitable consequence of war while quietly mourning the Menons:

Brazzoduro, a political maverick, arrived from Treviso and lodged with the Menon family, the owners of the automotive works – renowned in Italy and overseas as pioneers in the automobile industry. Brazzoduro, an inveterate Fascist, would be the downfall of would-be-victim Guglielmo Menon. Menon was a good man, a true Christian soul, but easily misled by the prospect of honour and power; which is why he, alongside friends and trusted employees, had founded the local branch of the Fascist Republican Party in Roncade. . . . He gave himself body and soul, always maintaining that he was doing it for all the right reasons. The birth of his Fascist entourage led to meetings, accusations and suspicions against everyone and everything even remotely associated with anti-Fascism.

(Citton, n.d.)

Menon would meet a gruesome end. Here is how the local sources interrogated by Favero explain the killing of the Menon brothers, a tragedy that shattered the dwelling practices of an entire community beyond political colours and loyalties in a truly destabilising event. A Partisan patrol led by the Ferretto, a Communist Spanish civil war veteran, passed through asking for monetary support for the Partisan cause on behalf of the Command. They entered the Menon farm and asked Guglielmo for money. When he refused, claiming to be 'on the other side', the Partisans

manhandled them [him and his brother] into a car loaded with muskets, hand grenades, rifles and machine guns, and they took them to an out-of-the-way place, where they were immobilized. . . . War is always cruel and often destroys innocent lives, and those relatively innocent. To distinguish between levels of culpability in the heat of battle, in the heat of the moment, is near impossible.

(Bizzi 2005: 101–102)

The last sentence in the above extract is intentionally ambiguous with regards to the guilt of all parties involved. Other locals gave their own version of events:

The four individuals who killed the Menons and their collaborator were led by Visentin, a native of the San Donà [di Piave] area. The others were a Partisan from the Mestre area and two [Southern] drifters, one from the Pugliese city of Foggia and a Sicilian. Their target was not that of getting money out of the Menons but to physically eliminate the brothers.[9]

When asked by Favero about the reason for the Menons' killing, a local veteran Fascist officer, P.M., answered:

Q: Why did the Menons' driver get away, of all people? [He had been an outspoken and needlessly cruel Fascist thug]
P.M.:   Probably because they didn't know him. Those Partisans who carried out the shooting were from Venice . . . from outside.

The most obvious explanation for this fiasco, he infers, is that the culprits had been outsiders (see Haslam et al. 2005). Only an outsider to local dwelling practices, no matter what their political convictions or how intense their hatred towards Fascism, would have executed the two men upon whom the entire local economy depended – fathers, as the local priest put it, to 300 factory workers. No local patriot would have deliberately plunged this rural community into chaos and instability through a political assassination.

Other elements complicate the picture of wartime experience in this little town. One particular category of otherness is fuelled by – as is often the case in northern Italian narratives of the Resistance – the inborn hostility towards other Italians: namely Fascists from the South. The worst stereotypes and attributes of the southern Other, a not insignificant part of the local dwelling perspective, populate Favero's stories (Favero 2003: 71):

> The composition of the *Brigate Nere* was . . . varied. You had local career-hungry officers as well as young recruits coming from the South wearing the same uniform, with the frequent addition of elements stemming from organized crime. With a few exceptions this lot was unscrupulous riff-raff, noisy and instinctive rabble that did not even bother to impress their German pals.

The Southern Fascists are seen as pathetic puppets to their militarily and disciplinarily superior German counterparts. One of Favero's sources, P.S., voices the opinion of many: 'In our area the worst elements were definitely the Fascists. The Germans, at least, had their code of conduct – the Fascists were psychopaths.'[10] In the Roncade story we recognise a sort of lingering awareness of the affective trauma felt by a local community. Between the lines we can almost experience for ourselves the tearing asunder of the local dwelling perspective – a long-established status quo. Civilian peace was torn by the assassination, again by an invasive, noxious Other, of the two main employers in the area. Still resented and commemorated today, this thoughtless act of anti-Fascist rebellion had truly devastating consequences for locals, and cast an unsavoury shadow on subsequent positive achievements by anti-Fascist patriots in the area. Roncade is one case in which local Fascism had lived peacefully alongside and within a non-Fascist community by virtue of the economic benefit and greater social good. If we wanted to interpret these micro-histories through the lens of infrahumanisation, we could argue that the community's dwelling practices were disrupted by the noxious actions of outsiders: except in this case the Other was a band of Partisans. A gash in the tapestry of the local affective geography had been torn, one that would linger in memory for a long time.

So, who *were* the enemies?

This chapter drafted some of the experiences of Italian individuals and communities blighted by civil war and foreign occupation during the Second World War.

It made a case for disruptions to local dwelling practices (signalled by a warped sense of place or feeling out of place) being brought forth by the alienating and destabilising Other, the infrahuman Other with whom one cannot quite understand or relate on a common emotional plane (as in Anna's interview). As can be learnt from memory narratives, positionings in relation to local dwelling perspectives determined who the enemy was and how one went about in their interactions with individuals or groups – the Other. In terms of identity formation and negotiations, the dehumanisation and othering of the enemy became the norm. Anthropologist Gregory seeks to understand the mundane cultural forms that 'mark other people as irredeemably "Other" and that license the unleashing of exemplary violence against them' (Gregory 2004: 16). This was certainly the case in occupied northern Italy. More importantly still, identity constructs and dynamics intrinsic in a dwelling perspective of the conflict are even more significant in its aftermath: by aftermath here I mean the purging actions and 'score-settlements' of wartime and pre-war events that often engendered internecine violence and post-war rifts in social and political life across Europe.

In this chapter, whenever possible, I have let the witnesses speak about their positioning vis-à-vis local dwelling practices in their own words. The use of micro-historical fragments rather than a paraphrasing of content seemed more conducive to a 'feel' for the emotional elements lurking between the cracks of words and pauses. It is unfortunate that a conventional book chapter could not convey the immediacy, the open-endedness and the uncertainty of speech. Part of a long oral tradition embedded in national and local dwelling perspectives, stories from the Italian front have been told, shared and relived. Some have been transcribed and some, as in this chapter, scrutinised through a more-than-representational analytics. Some stories stick, unshakeable; and some others are gladly forgotten. All bear witness to a time of unpredictable violence and danger in which familiar practices were disrupted beyond recognition.

The interweaving of personal, place specific and public memories and understandings of the war, even when expressed and worked through literary artifice (see Favero 2003) bears witness to an emotional continuity between the events narrated and the author's personal engagement with the living memory of the events. Dwelling perspectives often located at the margins of mainstream remembrances live on in the words and gestures of the witness storyteller – in the micro-histories of affect (De Nardi 2014a, 2014b). In our Italian case study, the infrahumanisation of the enemy Other disrupted familiar dwelling perspectives and practices that had been established prior to the German occupation.

Finally, we have seen how otherness was embedded in the foreign occupying forces, the Germans and in the Fascists who, as Italians from 'elsewhere', are still dramatically woven into Resistance narratives due to their penchant for cruelty. The German and Fascists' infrahuman otherness was exacerbated by their non-belonging in local dwelling perspectives; as infrahuman outsiders, incapable of uniquely human emotions such as mercy and regret (see Leyens et al. 2000), the enemies were woven into anti-Fascist dwelling perspective as out of place, anxious beings, and their fear was deemed to be unavoidable (as glimpsed in the interviews with Lorenzo and Giorgio). However, in Roncade infrahumanisation practices extended to the anti-Fascists when they acted as outsiders destabilising the local community's dwelling practices. These were Resistance fighters who acted and interacted with local elements in 'foreign', and potentially disrupting, ways. In Roncade, as in other post-war Italian towns (Cappelletto 2003), we encounter a sense of the 'the desire for peace in the majority of Italians [that] caused them to perceive the presence of Partisan bands as potentially threatening, whether or not they were reckless' (Pezzino 2005: 407). Sense of place, however irrational, unrealistic, dystopian and topsy-turvy it may be, played a vital role in the cultural and personal understandings of the dynamics of the war and Resistance episodes. The enemy was simultaneously both the confused and the confounding Other. The enemies, whatever their political colours, were perceived with the dizzying sense of disorientation that a strange, foreign place might inspire. The identity of the enemy, always blurred and often contradictory, depended wholly on local dwelling perspectives. If the experience of the war and Resistance are to be thought of as a web of dwelling practices made up of places, individuals and emotions shaped and influenced by what people did in the war and where they came from, then the enemy Other is the great unknown in this dwelling perspective – the infrahuman element throwing chaos into the whole affair.

Notes

1   Author's interview with Liviano Proia on 6 August 2011.
2   Interview by Claudio Pavan with A. Busatto on 21 January 2014. Interview and Italian transcript online at www.youtube.com/watch?v=wxNjnG2jV1g.
3   *Platzkommandatur* translates as 'German Headquarters'.
4   Interview by Claudio Pavan with C. Mazzucco. Interview and Italian transcript online at www.youtube.com/watch?v=wGFvtWTG0ZA.
5   Author's interview with Anna Granzotto on 13 August 2010.
6   Author's interview with Giorgio Vicchi on 20 December 2010.
7   Author's interview with Lorenzo Altoè on 18 December 2011.
8   www.anpitreviso.it/?page_id=579.
9   Author's interview with Gianni Favero on 8 May 2012.
10   Author's interview with Gianni Favero on 13 May 2012.

References

Ahmed, S. 2004. Collective feelings. Or, the impressions left by others. *Theory Culture Society* 21 (2): 25–42.

Ankersmit, F. 1996. Can we experience the past? In R. Torstendahl and I. Veit-Brause (eds) *History-Making: The Intellectual and Social Formation of a Discipline*. Stockholm: Almqvist and Wiksell International, 47–76.

Askins, K. and Pain, R. 2011. Contact zones: participation, materiality, and the messiness of interaction. *Environment and Planning D: Society and Space* 29 (5): 803–821.

Battaglia, R. 1953. *Storia della Resistenza Italiana*. Turin: Einaudi.

Behan, T. 2009. *The Italian Resistance. Fascists, Guerrillas and the Allies*. London: Pluto.

Bizzi, I. 2005. Il *cammino di un popolo. Antifascismo e resistenza dal Brenta al Tagliamento. 1924–1944*. Venice: Marsilio.

Brescacin, P. 2014. *Il Sangue che Abbiamo Dimenticato*. Vittorio Veneto: Tipse.

Cappelletto, F. 2003. Long-term memory of extreme events: from autobiography to history. *Journal of the Royal Anthropological Institute* 9: 241–260.

Citton, Romano. n.d. Note Riassuntive del Periodo Bellico. 1943–1944–1945. Unpublished notes.

De Nardi, S. 2014a. No one had asked me about that before: a focus on the body and 'Other' resistance experiences in Italian World War Two storytelling. *Oral History* 43 (1): 5–23.

De Nardi, S. 2014b. An embodied approach to Second World War storytelling mementoes: probing beyond the archival into the corporeality of memories of the resistance. *Journal of Material Culture* 19 (4): 443–464.

Favero, G. 2003. *Inesorabile Piombo Nemico*. Roncade: Edizioni Piazza. Available online: www.roncade.it/download/2008/Inesorabile%20Piombo%20Nemico.pdf, last accessed 26 April 2013.

Forlenza, R. 2012. Sacrificial memory and political legitimacy in postwar Italy. Reliving and remembering World War II. *History & Memory* 24 (2): 73–116.

Gregory, C. 2004. The oral epics of the women of the Dandakaranya Plateau: a preliminary mapping. *Journal of Social Sciences: Interdisciplinary Reflection of Contemporary Society* 8 (2): 93–104.

Haslam, N., Bain, P., Douge, L., Lee, M. and Bastian, B. 2005. More human than you: attributing humanness to self and others. *Journal of Personality and Social Psychology* 89 (6): 937–950.

Henig, D. 2012. Iron in the soil: living with military waste in Bosnia-Herzegovina. *Anthropology Today* 28 (1): 21–23.

Hirsch, M. 2008. The generation of postmemory. *Poetics Today* 29: 103–128.

Ingold, T. 1993. The temporality of landscape. *World Archaeology* 25 (2): 152–174.

Ingold, T. 2000. *The Perception of the Environment: Essays on Livelihood, Dwelling and Skill*. London: Routledge.

Jackson, A. and Mazzei, L. 2011. *Thinking with Theory in Qualitative Research: Viewing Data across Multiple Perspectives*. London: Routledge.

Lewis, L. 1985. *Echoes of Resistance. British Involvement with the Italian Partisans*. Tunbridge Wells: Costello.

Leyens, J. 2009. Retrospective and prospective thoughts about infrahumanization. *Group Processes & Intergroup Relations* 12 (6): 807–817.

Leyens, J., Paladino, P., Rodriguez-Torres, R., Vaes, J., Demoulin, S., Rodriguez-Perez, A. and Gaunt, R. 2000. The emotional side of prejudice: the attribution of secondary emotions to ingroups and outgroups. *Personality and Social Psychology Review* 4 (2): 186–197.

MacLure, M. 2008. Broken voices, dirty words: on the productive insufficiency of voice. In A. Jackson and L. Mazzei (eds) *Voice in Qualitative Inquiry: Challenging Conventional, Interpretive, and Critical Conceptions in Qualitative Research*. London: Routledge, 97–113.

Moshenska, G. 2008. A hard rain: children's shrapnel collections in the Second World War. *Journal of Material Culture* 13 (1): 107–125.

Navaro-Yashin, Y. 2012. *The Make-Believe Space: Affective Geography in a Postwar Polity*. Durham, NC: Duke University Press.

Nordstrom, C. 1997. *A Different Kind of War Story (The Ethnography of Political Violence)*. Philadelphia: University of Pennsylvania Press.

Paladino, P.M., Leyens, J., Rodriguez, R., Rodriguez, A., Gaunt, R. and Demoulin, S. 2002. Differential association of uniquely and non-uniquely human emotions to the ingroup and the outgroup. *Group Processes and Intergroup Relations* 5 (1): 105–117.

Pavone, C. 1991. *Una Guerra Civile. Saggio Storico Sulla Moralità della Resistenza.* Turin: Bollati Boringhieri.

Pezzino, P. 2005. The Italian Resistance between history and memory. *Journal of Modern Italian Studies* 10 (4): 396–412.

Radstone, S. 2007. Trauma theory: contexts, politics, ethics. *Paragraph* 30 (1): 9–29.

Reddy, W. 2001. *The Navigation of Feeling: A Framework for the History of Emotions.* Cambridge: Cambridge University Press.

Rosenwein, B. 2006. *Emotional Communities in the Early Middle Ages.* Ithaca, NY: Cornell University Press.

Saunders, N. (ed.) 2004. *Matters of Conflict. Material Culture, Memory and the First World War.* London: Routledge.

Saunders, N. 2009. People in objects: individuality and the quotidian in the material culture of the war. In C. White (ed.) *The Materiality of Individuality. Archaeological Studies of Individual Lives.* London: Springer, 37–55.

Spagnol, T. 2005. *Memoriette del Tempo Nero. A Cura di Pier Paolo Brescacin.* Vittorio Veneto: Tipse.

Thrift, N. 2004. Intensities of feeling: towards a spatial politics of affect. *Geografiska Annaler Series B – Human Geography* 86 (1): 57–78.

Valentine, G. and Sadgrove, J. 2012. Lived difference: an account of spatio-temporal processes of social differentiation. *Environment and Planning A* 44 (9): 2049–2063.

Vendramini, F. 1984. Note sul collaborazionismo bellunese durante l'occupazione tedesca (1943–1945). In F. Vendramini (ed.) *Tedeschi, Partigiani e Popolazioni nell'Alpenvorland.* Venice: Marsilio Editori, 171–208.

Wetherell, M. 2012. *Affect and Emotion. A New Social Science Understanding.* Milton Keynes: Open University Press.

7 Sound memory

A critical concept for researching memories of conflict and war

Carolyn Birdsall

In recent geographical literature, there has been an increased sensitivity to sound in urban space and everyday life, with particular interest in the intersections between music, identity and place (Revill 2000; Connell and Gibson 2003; Boland 2010). In the wake of non-representational theory, cultural geographers have paid additional attention to affect and embodiment, and to the role of the senses in remembering (Anderson 2004; Thrift 2004). This chapter contributes to the interest in the study of embodied and emotional geographies, in which emotion is understood terms of 'its socio-*spatial* mediation and articulation rather than as entirely interiorized subjective mental states' (Bondi et al. 2005: 3). Where my previous work evaluated the figure of the 'earwitness' for researching historical soundscapes (Birdsall 2009, 2012), the present chapter is centred on the concept of 'sound memory' as a tool for understanding performances of past war experiences in the present, and establishing these acts as (re)constructions of the self in relation to others and to place.

While non-representational theory was an important touchstone for my earlier investigations, this chapter's treatment of sound memory and the remembrance of war may be better described in terms of the more-than-representational (Lorimer 2005). In opting for 'more-than-' rather than 'non-', this analysis will maintain an interest in the affective, expressive and performative, without necessarily dismissing the role of social discourse and cultural practices in acculturating the body (Bondi 2005). Given this interest in both the affective and discursive, and the individual and social dimensions to remembering, I will develop the concept of sound memory with the aid of psychologist Pierre Janet's theorization of memory, narration and trauma. Janet's tripartite model distinguishes between *habit memory*, *narrative memory* and *traumatic memory* (Janet 1973). The first, habit memory, comprises the habitual bodily skills acquired over time, and through routine, whether unreflective or acknowledged by the subject. The second category, narrative memory, refers to the act of creating a story or description of past events. These memories are often intentionally given emotional significance, but can even encompass those memories that seem to emerge by chance. The final category described by Janet, traumatic memory, requires critical reflection on the specificity of trauma and modes of traumatic recall in interview situations.

Sound memory will be explored here as a tool for understanding the role of sounds – both past and present – in acts of remembering war in German cityscapes. The first section will establish the significance of sound in war and its remembrance, and reflect on my use of oral history methods to observe invocations of space, place and identity during interviews with elderly people about the sounds of war. The analysis will then take up the three categories developed by Janet, focusing first on habit memory, in terms of how interviewees recalled and performed sound-related aspects of their childhood socialization. Narrative memory will be analysed by the ways in which individual subjects framed their recollection of musical sound in terms of place-making practices (both terms of local attachments and national frameworks), before focusing on a group interview setting that drew on a familiar repertoire of musical song from their childhood years in which shared place and identity are enacted and positive aspects of their younger years were reclaimed. This performance of *in situ* belonging appeared to downplay internal differences and steer clear of unpleasant memories of the past. Finally, I will reflect on the places in which overwhelming sounds were experienced during protracted exposure to Second World War aerial attacks. This category requires critical reflection on different modes of traumatic recall and the extent to which the accounts studied here should be framed in terms of trauma. My engagement with Janet's model will work towards a final conclusion about being attuned to the affective and (inter)personal negotiations of memory during acts of remembering, and how they reveal telling articulations of place and identity.

Placing sound, conflict and the politics of memory

With the emergence of Sound Studies, with its particular interest in sound history, much attention has been paid to the role of sound in defining the self, belonging and identity (Bull 2000; Connor 2004; Born 2011). For the case of war and conflict, scholars have tended to emphasize the social functions of music in boosting morale and shaping cultural experience (Watkins 2003; Fauser 2013). At the same time, much sound historical scholarship has been devoted to the central role of sound (and noise) in conflict and warfare, whether in forms of social exclusion and marking of territories, in civil protest, and forms of battle and interrogation (Deaville 2012; Cusick 2013). While such patterns have been defined as transcultural and transhistorical (Schwartz 2011; Hendy 2013), scholarly research on sound history has shown a strong interest in the civil and world wars of the modern era (Smith 2001). The battlefield of the First World War has attracted particular attention due to the reduced visibility and ear-piercing sounds of mechanized warfare, and its marked effects on those who experienced it (Jean 2012). The bombing of Munich, Hamburg and Freiburg during the First World War also sparked extensive preparations during the 1920s and 1930s for the possibility of noisy aerial attacks on German cities (Geinitz 2000).

Among the methodologies explored in sound historical research, one approach that has remained underexplored is that of oral history. My main motivation for adopting this methodology is that the interview process encourages a mode of remembering that elucidates the sensory and embodied nature of experience (Hardy 2006), and processes of identity and place-making in relation to the remembrance of war. The oral history methodology allows for memories to be activated and constructed during the present moment of recall, facilitated by embodied and more-than-discursive forms of performance (Cándida-Smith 2002: 1–3).

Memories performed in interview contexts are equally sensitive to a variety of other factors, not least the choice of questions, location, language and mood of those involved (Perks and Thomson 2006). Indeed, all acts of memory involve a 'composite of truth and fiction' dependent on a present-day interpretation and social context of remembering (Hutton 1993: 64; Halbwachs 1992). In turn, recent oral history research has stressed that, while interviews may produce recollections with historical errors, such accounts not only offer insights into how past events are reencountered in the present, but also into the role of place and spatial strategies within narrations of (difficult) pasts (Cole 2015). The analysis that follows will thus acknowledge the presence of body language and gestures as comprising an integral part of the interview (as intersubjective encounter), and also consider in detail the ways in which place figured in interviewees' responses about experiencing National Socialism and the Second World War during their formative years.

The first significant perspective on place for my oral history project can be identified in the location of (most) interviews at the Stadtsarchiv Ratingen in the north of Düsseldorf. A full, front-page article in a daily regional newspaper, entitled 'Die Klangwelten des Krieges' ('The soundscapes of war'), invited potential interviewees to leave their details with the archive (Hartleb 2004a).[1] A colour photograph depicted me standing in the archive stacks with the director Erika Münster-Schröer, foregrounding both the site of the archive and legitimizing my project in the context of a familiar local institution. The interviews at the archive were conducted in a room (at ground level), which looked onto the playground of the Anne-Frank-Schule, a primary school behind the archive. The frequent school bell rings and sounds of children playing were commented on by interviewees, and appeared to serve as an auditory prompt for their processes of recalling childhood and young-adult experiences. Moreover, the present-day association of the archive building with National Socialism was commented on by several interviewees, since the building housed Düsseldorf's regional Gestapo headquarters in the final years of the Second World War (Kaminsky 1991).

Thirty individuals – born between the late 1910s and the 1930s – volunteered to be interviewed, and I attended two meetings of a local seniors group for women. In several exceptions, the interview was conducted in the interviewee's home, usually if they were unable to travel easily. In many of these cases, interviewees were more likely to act like a host during a social visit, and use memory aids (e.g. photos, songbooks) during interviews. A number of the respondents had very little experience in providing an extended account of their past, whereas a sizeable portion had previously produced forms of life-story narration (e.g. unpublished memoirs, media interviews and local history activities). Many interviewees expressed their curiosity in an 'intergenerational encounter' (Vanderbeck and Worth 2015) with someone identified as a young, foreign researcher, with several citing a concern about negative perceptions of Germans as additional motivation for their participation. More broadly, the interviews appeared to represent a form of 'making connections' that also entailed a therapeutic component (for putting things in their place) (Bondi 2005). At the same time, I was mindful of the ethical issues when talking to older people about personal experiences of conflict and war; as a process that, for many, was linked to emotionally charged memories.

While most participants were similarly born in lower- to middle-class households, it became apparent that their remembering of self and place revealed an important distinction between those who grew up (and have stayed) in the Düsseldorf area, and those who experienced displacement during or following the Second World War (e.g. the mass flight and forced migration of German populations from Central and Eastern Europe).

For my semi-structured interviews, the questions first concerned general biographical information (including questions about childhood, family, school and routines) and, second, memories of sound in everyday life (including radio and other audiovisual technologies). When making appointments to be interviewed, a number of participants asked for both the permission sheet (with vital information) and a preliminary list (both questions and keywords) to be posted ahead of the interview; this prompted several interviewees to (partly) read from typed answers, or show discomfort when my follow-up questions required them to speak freely. Such strategies might be read in terms of anxiety about accuracy, which is common among elderly interviewee subjects. Nonetheless, they also reflect a more widespread concern about accurate remembering of National Socialism: often on the basis of forgetfulness (given their present older age), but also anxiety about being subject to questions about accountability (particularly on the issue of whether parents and family members were party members); many pointed out that their (non-adult) experiences were primarily limited to localized sites of the home and street, and to school, religious and youth-group participation. Recent literature has suggested the value of such localized per-spectives on how older people understood the self, as produced in a relationship to others and place. Oral histories give insights

> into place attachments and identities on scales ranging from the country to the region, town, street and home [and] into contests over place involving control, resistance and negotiation. This is evidenced in our case study by the breaking of safety rules and regulations by children [in the Second World War].
>
> (Andrews et al. 2006: 170)

Such oral history interviews therefore reveal memories of place as constituted on interrelating scales (micro to macro geographies), as well as the particular conditions – as demonstrated above for the south-east English coastal town of Teignmouth – of children's geographies.

At the same time, interview respondents – from the vantage of the present – can reflect on how this past is located and has meaning in the contemporary situation. While some mentioned everyday social geographies of the present (e.g. not wanting to greet certain people in the street due to their actions during National Socialism), others talked about how their own processes of recalling place experiences were prompted by media reporting about Iraq during 2003–2004. One such illustration comes from Ursula S. (born 1928, personal interview):

A: When my mother told me stories [about starvation, influenza and losing relatives during the First World War], I couldn't imagine it. Just as it is for you – these stories are so impossible.

Q: Is it difficult to picture these things in your mind?

A: It's better if you try not to. You know, when I was watching TV when the Iraq War began, would you believe that it seemed like it was there again. I woke up screaming from my bed, because the bombs were falling again and the devil knows what else. . . . For that reason, I don't watch TV anymore.

Such comments are suggestive of how US aerial bombings and urban warfare in Baghdad and other cities served as a ubiquitous (and unwanted) audiovisual prompt for my interviewees when reconsidering their own experiences of military attack and urban destruction during the Second World War.

One other important discursive context at the time of my interviews was the growing acknowledgement of German civilian suffering during the Second World War, due to air attacks (Sebald 2003) or flight and expulsion from the East (Grass 2002). A study of expelled Germans (living near Hannover) argued that publications by novelists such as Günter Grass were part of a 'genuine attempt to link flight and expulsion to a comprehensive narrative that does not omit German atrocities and does not regard everyone just as victims' (Schulze 2006: 378). Other oral history studies in this period observed that intergenerational family narratives overwhelmingly centred on stories about aerial bombings, and frequently made use of popular-culture references as an interpretative framework (Tschuggnall and Welzer 2002). These latter findings were also subject to critique, given the tendency to focus on elderly interviewees as unreliable narrators, rather than consider the intersubjective processes by which researchers are 'co-constitutive of narrative meaning' (Fasulo 2002: 115). Such concerns also draw attention to the 'emotional dynamics' of research relationships (Bondi 2005), a theme to which I will return in the analysis that follows.

My interviews therefore not only reflect a temporal interval – with remembering filtered through their present situation and experiences since 1945 – but also the often-fraught politics of memory in post-Holocaust Germany. A remembrance process has been studied that emerged from the late 1940s onwards, in which public discourses have been characterized in terms of a 'selective remembering' rather than complete silence (Hughes 2000: 205). Processes of 'normalization' have been observed from the 1950s onwards (Niethammer 2001), and culminated in the *Historikerstreit* (Historians' Debate) in the 1980s, which centred on the place of the Holocaust in Germany's larger historical narrative (Habermas 1989; Niven 2002). Recent memory studies scholarship has called for attention to contemporary dynamics of 'multidirectional memory'; focusing less on questions of German guilt, but rather on how individuals and collectives negotiate their place as 'implicated subjects' (Rothberg 2013). In what follows, I will engage with the interview responses – using Janet's tripartite model of memory for considering sound memory – with a view to their invocations of identity and place-making in the past and present.

Sound memory as habit memory: embodied routines and spatial practices

The notions of embodied habit (and habitus) have offered important components for the theorization of memory (Bergson 1988; Mauss 1979; Bourdieu 1977). Habit memories, in Pierre Janet's understanding, involve a set of accumulated, sometimes unreflective memories based on routines or habit. In many cases, these memories are based on bodily, muscular acts that are easily recalled and performed, but involve a knowledge that is not actively acknowledged or verbalized. During remembering, habit memory involves moments where the past is reencountered in the present – by means of both sound-making and gesture – and play an important role in contributing to a sense of self and belonging (Casey 1987: 151). By drawing on Henri Lefebvre's work, habit will be positioned as a social(izing) process that is often reestablished during times of social-political transition, involving a historically specific process of repeating and internalizing particular actions. The centrality of National Socialist rituals to the habit memories analysed here also serves as a reminder of how the 'performativeness' of ritual connects memory practices with embodied experience, as well as the intertwining of different scales of identity and place (Edensor 2002).

According to Lefebvre, by teaching children how to repeat 'a certain act, a certain gesture or movement' (2004: 39), they adopt a series of corporeal habits and accepted values, such as posture, attention, mannerisms, codes of conduct and etiquette. This is a process of legitimation where the interaction of acceptable habits and a value system reinforces both elements. Lefebvre determines this process to be one of force, and gives it the French equestrian term '*dressage*'. This term clearly indicates that the apparatus by which children are socially trained and acculturated is by no means gentle, as they are 'broken in' like animals (Lefevre 2004: 39). Lefebvre's concept of the *dressage* process can be applied to the strong National Socialist influence in both the school environment and everyday lives of children (Michaud 1997). I employ *dressage* here to explain the physical positions (e.g. the *Heil Hitler* salute) and habits (e.g. rote-learning of historical dates and slogans) taught to children, as part of a process in which certain sounds and music were performed with codified physical actions in Nazi pedagogy.

Dressage in this context can be identified in terms of collective singing, marching and 'call-and-response' interactions, which took place on a regular basis and were reinforced in the weekly activities led by the *Hitler Jugend* organization. In addition to the active performance of rituals, the *dressage* experienced in schools and youth organizations involved compulsory uses of radio and film propaganda (Gauger 1943: 26). Interviewees cited the introduction of Hitler portraits into every classroom as an important component of these repeated rituals; nationalist narratives in classroom teaching intensified in the lead up to, and during, the war (Johnson 2008). Following Pillemar (2004: 150), this schooling in a specific repertoire of rituals can be read as participating in a 'collective knowledge', which he cites as a more appropriate term than collective memory to refer to community codes of conduct and memory processes. Interviewees frequently cited the renaming of prominent streets and squares after Hitler and other figures after 1933 (Kleinfeld 1996), as one of the ways that local spaces were invoked as part of the national project. The repeated use of these streets and squares by youth groups to for marching, singing and recitation   along with classrooms, schoolyards and youth-group centres – is further suggestive of how localized activities and civic service were framed as participation in the national project.

During my oral history interviews, many of the elderly respondents used physical gesture to demonstrate their experience of this dressage during their childhood. A number of interviewees recalled the repetitive or punitive nature of these routines:

> On the first day of the school year we had to attend a flag-raising ceremony. With an outstretched arm – that was not allowed to be propped up [by the other hand] – we sang *Deutschland über alles* and *Die Fahne hoch!* [the two national anthems]. Once, during a flag-raising ceremony, I had a bleeding wound on my face as I had run into a post box. But only after I had taken part in the ceremony did my teacher send me home.
>
> (Renate S., born 1928, personal interview)

> At the Catholic school we still had [morning] prayers. But here [at the new school], everyone gathered in the playground at 7.55am. And then, after a saying of the day (*Tagespruch*), we went into class. . . . Older kids would make a list of those who were late.
>
> (Hannelore H., born 1927, personal interview)

Overall, it was fairly common to condemn the restrictions and requirements demanded by the Nazi system, yet also express that they enjoyed the sense of camaraderie and group activities (e.g. handicraft in the case of girls' groups). It was clear that the *dressage* experienced during National Socialism remains firmly part of the respondents' corporeal or habit memory, whereby remembering the routines and sites of school experience invokes somatic responses. Most interviewees had difficulties in acknowledging how these everyday routines – that in their daily repetition – had provided a sense of orientation and formed part of the self. For instance, most interviewees said they did not like many aspects of their schooling and *Hitler Jugend* experiences, particularly militarism and intimidation. This ambivalence was more pronounced for the case of men who were involved in pre-military training, with a number going into great detail about how these experiences took place at the nearby Ratingen Stadion site (all mentioning how the youth centre is now the site of a Spanish cultural centre).

In a similar way, the performance of the *Heil Hitler* greeting was often framed in terms of an exploration of adult spatial boundaries in the context of the regime. One example is Margarethe S. (born 1931) who noted that parents were anxious about inappropriate behaviour in public, after an uncle was picked up by the Gestapo after refusing to give the *Heil Hitler* greeting when entering a local tram:

> Something similar happened with a female colleague of my father – who was a doctor – and lived a few houses away from us. One night a patient arrived and said '*Heil Hitler*', and she said 'Stop that nonsense.' Two hours later the Gestapo picked her up, and my parents said at dinner 'For God's sake children, don't say anything, don't say "*Gruss Gott*" or such things'. So one day, a friend and I walked down the main street [of Hildesheim] and said '*Gruss Gott*' to every person we passed. But the people didn't react and nothing happened at all. It was a kind of children's opposition or testing the limits. But still as children we sensed that there was a climate of fear.

Stories like this were often cited by interviewees as illustrative of how children explored the social norms of the regime, along with various explorations and appropriations of urban space in the context of war (e.g. collecting colourful bomb fragments, Allied leaflet propaganda and so on).

Within Lefebvre's concept of habit memory, the process of *dressage* – phrased in terms of coercion and force – has a certain deterministic quality. A more nuanced analysis of sound, performative acts and identity construction would not, like Lefebvre, necessarily view identity as something solely determined by the individual's surroundings. Taking a cue from theories of gender performativity, Jonathan Culler (2002: 513) suggests that identity is not merely a 'condition one enacts', but something that is also tested and played with. Musicologist Tia DeNora also suggests that a dynamic between cultural scripts and individual choices provides an essential alternative to deterministic concepts of social power that designate all behaviour and actions in a given setting as dictated by official powers. Her analysis suggests that social experiences of music are constituted both in terms of 'action-repertoires', which function like social scripts, and 'action-strategies' that allow for resistances or modifications (DeNora 2003: 118–149).

By taking up DeNora's approach, we can consider habit memory as a dynamic process involving the social conditioning of sensory perception and the responses of bodies enacted upon. In the case of the interviewees, not only had their bodily actions changed over time, but also their childhood identification with these routines was abruptly made taboo following the war, and is discursively reframed from the standpoint of the present. As such, while all these performances reflected unique encounters, some interviewees were more rehearsed than others in narrating this past and performing physical gesture in communicating vivid memories that took place in the Düsseldorf area and elsewhere. These reflections bring me to the second category, 'narrative memory', which may involve more negotiations between the individual and social in present-day acts of recall, and their relationship to emplacement and identity.

Sound memory as narrative memory: autobiography, collectivity, identity

Janet's second category of narrative memory is most commonly concerned with 'explicit' memories invoking specific experiences, events or concepts. Often characterized in terms of autobiographical memory, such narration is dependent on and sometimes distorted by the context of reflection. As cognitive musicologist Bob Snyder (2000: 72) notes, long-term (musical) memories are affected by their individual and shared cultural components, since they 'are constructed largely from other memories of aspects of music previously heard, and other knowledge and metaphorical experience connected with music in our minds'. In most cases, individual interviewees included narratives about experiences of singing, listening and speaking in their spoken accounts. Such narration has been described in terms of a 'deliberate use of music to recollect, reminisce or recreate the content or mood of an already defined memory' (Anderson 2004: 13). Such practices may not actually represent the past but they reflect the use of music to bring up the past for the purposes of a present moment. As a result, the category of narrative memory can also involve the social uses of sound memories as tools for controlling affective space, as well as for establishing or preserving a stable sense of group identity.

One predominant instance of narrative memory was the engagement with popular and folk songs, which formed an integral factor, as Philip Bohlman (2002) has shown, in consolidating local landscape with region, nation and empire, and formed a central part of the National Socialist cultural programme and social life during the 1930s and 1940s. The strong identification with song repertoire of their youth became evident as a number of (female) interviewees told me that they hid their *Bund Deutscher Mädel* (BDM) song book *Wir Mädels singen* after the war to prevent their confiscation; these song books were described as treasured possessions, despite interviewee criticisms of the National Socialist regime (see Brade 1999).

A number of interviewees brought their songbooks to the archive to show me the new songs and lyrics that they had to add to the book during the course of the Second World War. Along with personal photos, Leon E. (born 1929) brought his sister's songbook to his interview at the archive, as a means of pointing out the various song repertoires, particularly those introduced during the war:

> There was a different book for the boys. But everything [from our house] was burnt, so we didn't have these belongings anymore. Here you can see that the

teacher wrote down the order of the songs to be learnt for the flag-hoisting in primary school.

The songbook provided a memory aid, which also prompted Leon to recall how his family lost their home and belongings during the large attack on Ratingen on 22 March 1945 (which buried his grandfather). The book is one of the few posses-sions the family kept, but he also showed how he and his sister – after 1945 – had expressed frustration by defacing the book, crossing out passages and scrawling the word '*Nationalsozialismus*'. Renate S. (born 1929) also explained her strong attachment to the book, explaining that I could photocopy its pages, but she wouldn't leave it at the archive; after 1945 she feared it would be taken by Allied Forces, but considered herself lucky that their house was not searched. Similarly, Theresa B. (born 1925) left her book behind for me to read and return, but then phoned that she was too anxious about losing it and wanted to pick it up again from the archive. These examples illustrate how the mixed feelings about the song repertoire of National Socialist pedagogy were attached to BDM songbooks; these books had an ambivalent status as both comforting and prohibited 'souvenirs', which symbolized for most a concrete connection to the (sometimes lost) places from their childhoods.

While individual participants frequently sang the lyrics of well-known songs during interviews, the context of the women's group meeting offers a vivid example of the social process of negotiating sound memories (Birdsall 2009). These women had a usual routine for their monthly meeting at a community centre: first coming together to sing (and occasionally listen to a talk), followed by a communal lunch. The group singing largely drew on traditional songs, with one participant explaining during a break that 'we sang these as children: they are folk songs (*Volkslieder*) and still sung today' (Helga S., born 1927). She and other participants went on to list the most well-known spring, hiking and morning songs, many of which were associated with the *Bund* youth movements that pre-dated National Socialism.

When I accepted an invitation to this meeting, I had initially understood that I would talk separately to individual members. On arrival, the group was quite large and there was an expectation that I would speak about my home country, which I agreed to do during a subsequent visit. In addition to this expec-tation of reciprocity, I also realized that my attendance provoked a slight distur-bance, perhaps due to my status as an outside guest or due to the sound-recording equipment and my topic of research. During the group singing of traditional songs (mainly about togetherness and community), none of the members resisted the choice of songs, nor verbally intervened when one member gave an opening cue for a song. Following this, the organizer seized my pile of copies (with the preliminary questions I used for individual interviews) and handed one to each participant. While a small group sat with me and all began speaking at once, others appeared to show a certain reluctance in writing about their memories (i.e. experiences during National Socialism and wartime, rather than more broadly about music and singing).

Two important distinctions can be observed about the womens' backgrounds: first, roughly half were from the Düsseldorf area, whereas most of the others came from Central and Eastern Europe (and one participant was born in the Dutch East Indies in 1920) and settled in West Germany in the early post-war period. This latter group provided very little information in their written responses, often giving writing evasive answers (in 'yes/no' form) in response to specific questions about home, schooling and other sites of past experience. The second, noticeable pattern was that roughly half were born before the National Socialist takeover (between the ages of 7 and 13 in 1933, and between 19 and 24 in 1945) and the other half were born after the takeover (and were between the ages of 3 and 13 in 1945). In the written comments, it is noticeable that the older women included 'fond' memories of the period between roughly 1925–1939, whereas a number of the younger women either struggled to remember details about National Socialism, or cited post-war events as stronger in their memory (e.g. a radio announcement about the foundation of Israel in 1948). As the meeting drew to a close, one of the women prompted the group to informally break into light-hearted singing of 'Danke (für diesen guten Morgen)', a religious song that gained pop-song status in the 1960s (Bubmann 2010).

All of these participants were in a later stage of life, with even those who were not originally from the Düsseldorf area having lived there for the majority of their lives (at least 40 or 50 years); the format of group singing allowed for a light atmosphere that facilitated a sense of shared place. As such, they were involved in the active creation of a group sound memory in the present, which drew on a shared cultural background as a means of performing – through group song – a positive a shared feeling about the past and sense of belonging. Rather than dwell on difficult memories or various forms of internal difference (e.g. accents, age group, place of birth, or even class differences), their musical participation in the group indicated the sense of consolidating shared moods and feelings based on a familiar cultural repertoire.

This group process of remembering reinvokes DeNora's dynamic between 'action-repertoires' and 'action-strategies' that I introduced in the previous section, since there is a use of social scripts that are reworked in a present-day group situation. Indeed, the value of this discussion enables my study of sound memories to encompass both the creative processes of individual life stories and group remembering contexts, alongside the embeddedness of corporeal memories. Beyond the relationship of embodied and social practices of remembering the past – and related invocations of selfhood, intersubjectivity and place – the last aspect that needs further elaboration and qualification is Janet's final category of traumatic memory.

Sound memory as traumatic recall: anxiety, triggers, acting-out

The final category described by Janet, traumatic memory, is perhaps the most contested, and is usually associated with painful events, often those that pose a challenge to narrative integration; in much trauma theory, the patient has been posited as unable to narrate or integrate extreme or painful episodes into existing schemes of meaning. The notion that there is a lack of control, where the past 'spills' into the present (Caruth 1995; van der Kolk et al. 1996) has since been disputed, with numerous scholars arguing against this influential understanding of traumatic recall as involving an exact performance of traumatic memory without alteration (Leys 2000: 252; LaCapra 2001, 2004). In many cases of traumatic memory, as in the case of so-called 'flashbulb' memories, those affected are able to recall past experiences in considerable detail (Brown and Kulik 1977; Schachter 1995, 1996). It is important to be specific about what aspects of a testimony refer to traumatic experience, and to keep in mind that not everyone is traumatized after being exposed to (potentially) traumatizing experiences; it is thus crucial to differentiate between different types of trauma (Drozdzewski 2015: 2).

With this in mind, I will reflect on the acts of remembering by interviewees with long-term exposure to war and military attacks, and how these instances involve positionings of self and others, in relation to place. In discussing experiences of the air war, interviewees clearly engaged in a vivid remembering of the sites and physical spaces in which they experienced sirens and air attacks. The family home was a primary locus, and these narratives invoke the vulnerability of both the home and outdoor spaces. Home life was described as marked by the absence of fathers and other relatives due to military service, death or detainment. Domestic spaces were described as interrupted and permeable as a result of the compulsory billeting of soldiers and extensive observation by block wardens, who checked that individual apartments had darkened windows and cleared-out attic spaces with sandbags, fire swatters and buckets of water.

The intensification of aerial attacks from 1940 onwards were described as the most significant threat to the safety of home, symbolized by unpredictable air sirens and explosions, as well as strong smells, summarized by one of the seniors' group members as a 'musty basement smell, concrete dust, fire, ash, carbide waste'. Along with such smells, interview participant Leon E. (born 1929) recalled that:

> In the evenings, the sky was blood-red as the cities burned. Since Ratingen is near Düsseldorf, Duisburg, Mühlheim, Oberhausen and Essen, you could always see where the cities were burning. And during the day, when the smoke clouds rose high up, ash would fall over Ratingen.

With such observations, interviewees drew attention to how they tried to make sense of their own situation by listening to sounds outside their homes; the magnitude of aerial attacks also cued via smell and strong visual impressions in the vicinity of the home and local area.

The other places that interviewees reflected on in their accounts were basement air shelters, which – in apartment buildings – were shared with other neighbours or visitors. Along with official bunkers, the confined space of basements was the predominant site in which interviewees recalled experiencing sirens and attacks, in which overwhelming sounds were associated with darkness, restricted air and high levels of anxiety. For instance, as one interview participant recalled:

> The cellar walls shook when a bomb fell nearby. The worst was when the lights suddenly went out and we couldn't see anything. It was an oppressive feeling, the women, in particular, cried. . . . It was unnerving to sit there in the cellar. First it would get really quiet. Then you would hear a very light, constant humming sound. Then it would be dead silent, until the bombs started hammering down. That was horrible. We would all kneel on the floor and pray. My mother always prayed the loudest. One had the feeling that the cellar floors were rising up.
>
> (Charlotte S., born 1930, personal interview)

Taking shelter in basements was thus described in terms of both at risk (being underground in the event of a bombing) and enduring prolonged periods of uncertainty and boredom, before returning to bed or other activities. For those who used public bunkers during night-time attacks, an additional concern was expressed about whether one could get dressed and walk through dark streets in time to reach such locations.

The sense of being at risk while travelling to and from school was mentioned frequently, along with strong memories of spotlights and shooting by anti-aircraft stations. In this case, male interviewees tended towards enthusiastic narration of how the early stages of the war appeared exciting, describing their fascination with spotlights and the shooting down of enemy planes, along with how they explored bomb debris and swapped colourful pieces (*Bombensplitter*) at school. A number of women noted that they would become emotional upon seeing (or hearing about) enemy planes being shot down, for instance:

> My memories of airplanes are terrible. Before the war they were not an issue, and we didn't live far from the airport. Later, they were the things that robbed me of my rest at night, that destroyed our apartments, that made both my school and play-areas unusable, and that forced me to look for bunkers during emergency alarms. And they were the things that made me tremble when they were caught by spotlights of the anti-aircraft cannons, and about which I cried, since there were people inside when they were shot down. The air protection warden [in the public bunker] reprimanded me about this. . . . But I couldn't handle it when I saw planes go down. I thought, 'My God, there are people in there' – and then it would be shot down. One cannot describe this feeling.
>
> (Renate S., born 1929)

Recollections of attacks at school were often accompanied by discussions of school air shelters, which were then linked to discussions of frequent disruptions to schooling as classes were closed, relocated or evacuated to rural areas (*Kinderlandverschickung*), through which a number of interviewees recalled being geographically separated from and worried about their families. A particular flash point for Hans H. (born 1934) was a memory of a local teacher who was responsible for delivering death notices in Ratingen; he noted his own regret at telling his neighbour that that teacher had called for her, as 'when [this teacher] went around, people would know what it meant'.

Such remembering also served as a prompt for reflections on the vividness of these memories and the occasional concern about being overly emotional: 'I hope I didn't talk too much. It all came back to me because of these questions. I can really see it before me now: the beds in air-raid shelters (*Luftschutzbetten*) and so many other things' (Jenny E., born 1929). For one of my youngest interviewees, Hans M. (born 1938), sirens figured as a flash point for recalling several incidents in which bombs fell close to their home: 'Even today I can still hear the sirens.' After describing living permanently in the cellar in the last war months, Renate S. (born 1929) paused momentarily and then observed that 'these are all those situations that only occur to you again when you open the cupboard [of your memories]'. When talking about this last part of the war, interviewees recalled that they became increasingly aware of the failure of the war effort, with a number of interviewees described how – with westerly winds – it was possible to hear fighting in the Eifel region from late 1944. In the case of Ratingen, memories of the final period of the war usually refer to the severe bombing attack on 22 March 1945, during a period of six weeks of artillery fire prior to the arrival of US tanks and troops. A number of interviewees recalled their fear that active party members in their local neighbourhood would stage solo attacks upon the arrival of Allied Forces, whereas one of the youngest respondents (Hans M., born 1938) noted that he associates the end of the war with a memory of the rattling sounds of chain tracks, as the military tanks rolled into town.

During the interviews, there were several cases indicating that the sounds of alarms and bomb explosions in the present were a cause for inducing fright and shock, due to the acoustic similarities felt between present-day alarms and the Second World War air-raid sirens. For instance, Therese B. (born 1925), who noted that she had rarely spoken at length about the events of her childhood and adolescent years, became increasingly emotional when interviewed. She explained that, upon hearing alarms or emergency sirens in the present day she experiences panic and has to hold her chest, before reminding herself that it is no longer wartime. The experiences of feeling overwhelmed in the present described here, among other aspects of Therese B.'s behaviour, appeared to be a strong indicator for an ongoing struggle to contain herself when hearing present-day sounds of sirens or loud planes. In such cases, the sirens create a trigger for the traumatizing sounds heard during childhood, and the ensuing panic experience can disrupt past–present distinctions.

A related instance occurred with the use of vocal sound effects in the moments when interviewees could not easily describe or narrate the felt experience of wartime aerial attacks. When talking about their memories of attacks on German cities, most interview participants were able to describe the sirens and safety precautions taken for air raids, but often could not describe the actual event of

bombings. In contrast to the description of routines and preparations, many interviewees used their arms to act out the flying over of planes and used their voices
to mimic the sounds of gunfire ('ra-ta-ta-ta-ta') or bombs exploding, as though
occurring in the present moment. Mimicking the sounds of gunfire was, in some
cases, linked to individual experiences of being shot at by low-flying aeroplanes,
as in the case of Leon E. (born 1929), Renate S. (born 1929) and Charlotte S.
(born 1930). Other imitations of sound were used to refer to radio fanfare
('dadadada' or 'tic tic tic'), signal jamming ('chk chk chk') or the sound of illegal
BBC radio ('boom boom boom'). The use of the present tense to enact the actual
event of the bombing was sometimes performed as occurring in the present. Such
vivid recollections also test the efficacy of language for describing the auditory
and sensory inscriptions of the repeated exposure to aerial attacks.

In recent years, the psychoanalytic conceptualization of trauma, repression and
working through has been criticized as an insufficient explanatory framework for
cultural or national memory (Langenbacher 2003), while other scholars have
similarly expressed doubt about whether the loss of community should be framed
as trauma (LaCapra 2004: 106–142). Nonetheless, as sociologist Kai Erikson
(1995) has argued, it is imperative to acknowledge that trauma can be generated
by isolated and sudden events, as well as over a sustained or prolonged period.
Erikson's understanding of collective trauma concerns how, when community ties
are severed, individuals are subject to a 'gradual realization that the community
no longer exists as an effective source of support and that an important part of the
self has disappeared' (1995: 187). Research on the specific role attributed to
noise and explosions for children exposed to aerial bombings, along with the
aftermath of these attacks, may support Erikson's expanded definition of trauma
(Somasundaram 1996: 1466, 1470; Gibson 1989). Such studies show a striking
correspondence with the intense bombings of the Düsseldorf area, which were
highly unpredictable, involved a complete overwhelming of the senses, and
occurred over an extended period.

Indeed, the most striking recollections discussed here were those concerned
with the sounds of sirens, planes and the attacks, with present-day triggers having
the potential to elicit anxiety or panic. In this case, at least, it did not seem to
be the case that these interviewees have had long-lasting traumatic symptoms.
Most interview participants were able to provide a fairly coherent narrative about
their life experiences and their volunteering to participate in my project suggests
that they were not avoiding these experiences. While prior engagement in memoir-
writing or other life-story activities might indicate well-rehearsed narration, the
interview situation often still elicited affective responses or emotional expressions
when discussing childhood experiences of war. As discussed above, the narration
of overwhelming events and the complete breakdown of social order involves
recollections about the self and the family (and community) that are inflected by
questions of physical space and place.

This chapter has elaborated on sound memory as a concept to capture both
the specific memories of sounds and the incidental sounds performed during the
process of remembering. In studying the embodied and interpersonal qualities of

the interview context, I have used Janet's theoretical framework as a means of teasing out different aspects of how remembering in interviews depends on physical expression and sound-making. Since most of the interviewees were children or young adults during the period in question, I have focused on how they narrated their experiences in terms of the home and neighbourhood, the educational system and National Socialist youth organizations. Such a task should not detract attention from Holocaust survivors and other victims of National Socialism and their acts of testimony (Morris 2001). In examining how interview participants negotiated difficult memories of the past, I have drawn attention to changing cultural norms that increasingly constructed Germans as victims, in view of air attacks, mass flight and forced migration. As the analysis has shown, this remembering process is in a constant state of renegotiation, influenced by both discourses from the regime period and the intervening years, along with the specific social and spatial dynamics of remembering National Socialism in the present.

Note

1   After the initial article, a follow-up was published six months later (Hartleb 2004b). The interviews were recorded with a minidisc recorder with a lapel microphone. The average birth year of individual interviewees was 1928, with an average age of 11 when the Second World War began in 1939.

References

Anderson, B. 2004. Recorded music and practices of remembering. *Social and Cultural Geography* 5 (1): 3–20.

Andrews, G.J., Kearns, R.A., Kontos, P. and Wilson, V. 2006. 'Their finest hour': older people, oral histories and the historical geography of social life. *Social and Cultural Geography* 7 (2): 153–179.

Bergson, H. 1988. *Matter and Memory*. Trans. N.M. Paul and S. Palmer. New York: Zone Books.

Birdsall, C. 2009. Earwitnessing: sound memories of the Nazi period. In K. Bijsterveld and J. van Dijck (eds) *Sound Souvenirs: Audio Technologies, Memory and Cultural Practices*. Amsterdam: Amsterdam University Press, 169–181.

Birdsall, C. 2012. *Nazi Soundscapes: Sound, Technology and Urban Space in Germany, 1933–1945*. Amsterdam: Amsterdam University Press.

Bohlman, P.V. 2002. Landscape–region–nation–Reich: German folk song in the nexus of national identity. In C. Applegate and P. Potter (eds) *Music and German National Identity*. Chicago: University of Chicago Press, 105–127.

Boland, P. 2010. Sonic geography, place and race in the formation of local identity: Liverpool and scousers. *Geografiska Annaler: Series B, Human Geography* 92: 1–22.

Bondi, L. 2005. Making connections and thinking through emotions: between geography and psychotherapy. *Transactions of the Institute of British Geographers* 30 (4): 433–448.

Bondi, L. Davidson, J. and Smith, M. 2005. Introduction: geography's 'emotional turn'. In J. Davidson, M. Smith and L. Bondi (eds) *Emotional Geographies*. Aldershot: Ashgate, 1–16.

Born, G. 2011. Music and the materialization of identities. *Journal of Material Culture* 16 (4): 376–388.

Bourdieu, P. 1977. *Outline of a Theory of Practice*. Trans. R. Nice. Cambridge: Cambridge University Press.

Brade, A.-C. 1999. BDM-Identität zwischen Kampflied und Wiegenlied: eine Betrachtung des Repertoires im BDM-Liederbuch *Wir Mädel singen*. In G. Niedhardt and G. Broderick (eds) *Lieder in Politik und Alltag des Nationalsozialismus*. Frankfurt am Main: Peter Lang, 149–165.

Brown, R. and Kulik, J. 1977. Flashbulb memories. *Cognition* 5 (1): 73–99.

Bubmann, P. 2010. Das 'Neue Geistliche Lied' als Ausdrucksmedium religiöser Milieus. *Zeithistorische Forschungen/Studies in Contemporary History* 7 (3): 460–468.

Bull, M. 2000. *Sounding out the City: Personal Stereos and the Management of Everyday Life*. Oxford: Berg.

Cándida-Smith, R. 2002. Introduction: performing the archive. In R. Cándida-Smith (ed.) *Art and the Performance of Memory: Sounds and Gestures of Recollection*. London and New York: Routledge, 1–12.

Caruth, C. 1995. *Unclaimed Experience: Trauma, Narrative, and History*. Baltimore: John Hopkins University Press.

Casey, E.S. 1987. *Remembering: A Phenomenological Study*. Bloomington and Indiapolis: Indiana University Press.

Cole, T. 2015. (Re)Placing the past: spatial strategies of retelling difficult stories. *Oral History* 42 (1): 30–49.

Connell, J. and Gibson, C. 2003. *Soundtracks: Popular Music, Identity and Place*. London: Routledge.

Connor, S. 2004. Sound and the self. In M.M. Smith (ed.) *Hearing History: A Reader*. Athens: University of Georgia Press, 54–66.

Culler, J. 2002. Philosophy and literature: the fortunes of the performative. *Poetics Today* 21 (3): 503–519.

Cusick, S.G. 2013. Toward an acoustemology of detention in the 'Global War on Terror'. In G. Born (ed.) *Music, Sound and Space: Transformations of Public and Private Experience*. Cambridge: Cambridge University Press, 275–291.

Deaville, J. 2012. The envoicing of protest: occupying television news through sound and music. *Journal of Sonic Studies* 3 (1): n.p.

DeNora, T. 2003. *After Adorno: Rethinking Music Sociology*. Cambridge: Cambridge University Press.

Drozdzewski, D. 2015. Retrospective reflexivity: the residual and subliminal repercussions of researching war. *Emotion, Space and Society* 17: 30–36. Available online: www.sciencedirect.com/science/article/pii/S1755458615000158, last accessed 4 February 2016.

Edensor, T. 2002. *National Identity, Popular Culture and Everyday Life*. Oxford and New York: Berg.

Erikson, K. 1995. Notes on trauma and community. In C. Caruth (ed.) *Trauma: Explorations in Memory*. Baltimore: John Hopkins University Press, 183–199.

Fasulo, A. 2002. 'Hiding on a Glass Roof', or a commentator's exercise on 'Rewriting Memories'. *Culture and Pyschology* 8: 146–152.

Fauser, A. 2013. *Sounds of War: Music in the United States during World War II*. New York: Oxford University Press.

Gauger, K. (ed.) 1943. *Bestimmungen über Film und Bild in Wissenschaft und Unterricht*. 4th edn. Stuttgart and Berlin: Kohlhammer.

Geinitz, C. 2000. The first air war against noncombatants: strategic bombing of German cities in World War I. In R. Chickering and S. Förster (eds) *Great War, Total War: Combat and Mobilisation on the Western Front, 1914–1918*. Cambridge: Cambridge University Press, 207–226.

Gibson, K. 1989. Children in political violence. *Social Science and Medicine* 28 (7): 659–667.

Grass, G. 2002. *Im Krebsgang*. Göttingen: Steidl.

Habermas, J. 1989. *The New Conservatism: Cultural Criticism and the Historians' Debate*. Trans. S. Weber Nicholsen. Oxford: Polity Press.

Halbwachs, M. 1992. *On Collective Memory*. Chicago: University of Chicago Press.

Hardy, C. 2006. Authoring in sound: aural history, radio and the digital revolution. In R. Perks and A. Thomson (eds) *The Oral History Reader*. 2nd edn. London: Routledge, 393–405.

Hartleb, R. 2004a. Die Klangwelten des Krieges. *Rheinische Post*, 27 March.

Hartleb, R. 2004b. Die Klangwelten des Bombenkrieges. *Rheinische Post*, 10 September.

Hendy, D. 2013. *Noise: AH History of Sound and Listening*. London: Profile.

Hughes, M.L. 2000. 'Through no fault of our own': West Germans remember their war losses. *German History* 2 (18): 193–213.

Hutton, P.H. 1993. *History as an Art of Memory*. Hanover, NH: University Press of New England.

Janet, P. [1889] 1973. *L'Automatisme Psychologique*. Paris: Société Pierre Janet.

Jean, Y. 2012. The sonic mindedness of the Great War: viewing history through auditory lenses. In F. Feiereisen and A.M. Hill (eds) *Germany in the Loud Twentieth Century: An Introduction*. Oxford and New York: Oxford University Press, 51–62.

Johnson, E.J. 2008. Under ideological fire: illustrated wartime propaganda for children. In E. Goodenough and A. Immel (eds) *Under Fire: Childhood in the Shadow of War*. Detroit: Wayne State University Press, 59–76.

Kaminsky, U. 1991. Die Gestapo in Ratingen 1933–1945: Verfolgungsprogramm und Verfolgungsrealität am Beispiel der Staatspolizeileitstelle Düsseldorf. *Ratinger Forum* 2 136–163.

Kleinfeld, H. 1996. *Düsseldorfs Strassen und ihre Benennungen von der Stadtgründung bis zur Gegenwart*. Düsseldorf: Grupello Verlag.

LaCapra, D. 2001. *Writing History, Writing Trauma*. Baltimore: John Hopkins University Press.

LaCapra, D. 2004. *History in Transit: Experience, Identity, Critical Theory*. Ithaca, NY and London: Cornell University Press.

Langenbacher, E. 2003. Changing memory regimes in contemporary Germany? *German Politics and Society* 21 (2): 46–68.

Lefebvre, H. [1992] 2004. *Rhythmanalysis: Space, Time and Everyday Life*. Trans. S. Elden and G. Moore. London and New York: Continuum.

Leys, R. 2000. *Trauma: A Genealogy*. Chicago: University of Chicago Press.

Lorimer, H. 2005. Cultural geography: the busyness of being 'more–than–representational'. *Progress in Human Geography* 29 (1): 83–94.

Mauss, M. 1979. Body techniques. In *Sociology and Psychology: Essays*. Trans. B. Brewster. London: Kegan Paul, 95–123.

Michaud, E. 1997. Soldiers of an idea: young people under the Third Reich. In G. Levi and J.-C. Schmitt (eds) *A History of Young People in the West*. Vol. 2: Stormy Evolution to Modern Times. Cambridge, MA: Belknap Press, 257–280.

Morris, L. 2001. The sound of memory. *The German Quarterly* 74 (4): 368–378.

Niethammer, L. 2001. 'Normalization' in the West: traces of memory leading back into the 1950s. In H. Schissler (ed.) *The Miracle Years: A Cultural History of West Germany: 1949–1968*. Princeton, NJ: Princeton University Press, 237–265.

Niven, B. 2002. *Facing the Nazi Past: United Germany and the Legacy of the Third Reich*. London: Routledge.

Perks, R. and Thomson, A. (eds) 2006. *The Oral History Reader*. 2nd edn. London: Routledge.

Pillemar, D.B. 2004. Can the psychology of memory enrich historical analyses of trauma? *History and Memory* 16 (2): 140–154.

Revill, G. 2000. Music and the politics of sound: nationalism, citizenship and auditory space. *Environment and Planning D* 18 (5): 597–613.

Rothberg, M. 2013. Multidirectional memory and the implicated subject: on Sebald and Kentridge. In Liedeke Plate and A. Smelik (eds) *Performing Memory in Art and Popular Culture*. New York: Routledge, 39–58.

Schachter, D.L. 1995. *Memory Distortion: How Minds, Brains, and Societies Reconstruct the Past*. Cambridge, MA: Harvard University Press.

Schachter, D.L. 1996. *Searching for Memory: The Brain, the Mind, and the Past*. New York: BasicBooks.

Schulze, R. 2006. The politics of memory: flight and expulsion of German populations after the Second World War and German collective memory. *National Identities* 8 (4): 367–382.

Schwartz, H. 2011. *Making Noise: From Babel to the Big Bang and Beyond*. New York: Zone Books.

Sebald, W.G. 2003. *On the Natural History of Destruction: With Essays on Alfred Andersch, Jean Amery and Peter Weiss*. Trans. A. Bell. London: Penguin.

Smith, M.M. 2001. *Listening to Nineteenth-Century America*. Chapel Hill: University of North Carolina Press.

Snyder, B. 2000. *Music and Memory: An Introduction*. Cambridge, MA: MIT Press.

Somasundaram, D.J. 1996. Post-traumatic responses to aerial bombing. *Social Science and Medicine* 42 (II): 1465–1471.

Thrift, N. 2004. Intensities of feeling: towards a spatial politics of affect. *Geografiska Annaler* 86 (B): 57–78.

Tschuggnall, K. and Welzer, H. 2002. Rewriting memories: family recollections of the National Socialist past in Germany. *Culture and Psychology* 8: 130–145.

Vanderbeck, R. and N. Worth (eds) 2015. *Intergenerational Space*. Abingdon and New York: Routledge.

Van der Kolk, B.A., McFarlane, A.C. and Weisaeth, L. (eds) 1996. *Traumatic Stress: The Effects of Overwhelming Experience on Mind, Body, and Society*. New York: The Guildford Press.

Watkins, G.E. 2003. *Proof through the Night: Music and the Great War*. Berkeley: University of California Press.

8   Heralding Jericho

Narratives of remembrance, reclamation and Republican identity in Belfast, Northern Ireland

Lia Dong Shimada

At 1,562 feet above sea level, rising from the western edge of the city, Divis Mountain is the highest point in greater Belfast. On a clear day, its summit offers sweeping views of Northern Ireland. In its time, Divis has served as the burial ground for ancient Irish kings; the source of the rivers that fuelled Belfast's thriving nineteenth-century linen industry; and, perhaps most controversial in recent memory, a major twentieth-century operational base for the British army. Today, Divis Mountain can be read as a contested landscape that illuminates the fraught history of Northern Ireland.

The long-standing conflict has been characterized by a high degree of violence on both sides of a deeply entrenched sectarian divide. At stake was – and is – the political identity of the six counties that comprise Northern Ireland. Broadly speaking, the conflict pits predominantly Catholic nationalists (who believe that these counties should belong to the Republic of Ireland) against predominantly Protestant Unionists (who advocate maintaining the region's current membership of the United Kingdom of Great Britain and Northern Ireland). A period of civil unrest, euphemistically known as 'the Troubles', erupted in 1969 and unleashed three decades of turmoil across Northern Ireland and beyond. On the nationalist side, groups such as the Irish Republican Army (IRA) took up arms for a united Ireland, while forces loyal to the United Kingdom fought to maintain the status quo. Years of negotiation led to the signing of peace accords in 1998. By this time, the conflict had claimed more than 3,000 lives (many of them civilian), entrenched patterns of segregation, and left a deeply traumatized society in its wake.

Over the past two decades, as peace has settled in Northern Ireland, Divis Mountain has undergone dramatic transformations. In the wake of the peace accords, the British military released its holdings on Divis, leading to the mountain's acquisition for and by the public. Underpinning the process of demilitarization, however, are multiple, contested readings that complicate the mountain's symbolic presence in 'post-conflict' Northern Ireland. In this chapter, I explore the challenge of reworking notions of culture, heritage and identity in a place recovering from violent conflict. The transformation of Divis – from militarized terrain to public open space – provokes important questions about the diverse, shifting expressions of cultural identity that are emerging through the mountain's wartime shadows. These are apparent in the ongoing tensions between anti-imperial narratives in contemporary Irish Republican culture, and in the need for new and wider ways of conceptualizing the mountain for a time of peace.

I situate my analyses within the rich seam of cultural geography, with its theoretical rigour in defining 'landscape'. For example, Crang (1998: 162) sees cultural landscapes as agents in the process of imbuing territory with the ideas of a specific cultural identity. Along similar lines, Hardesty (2000: 171) argues that people invoke their own cultural and social images in the creation of cultural landscapes that both reflect and form the continuous process of 'world-making'. I draw particularly on David Matless' (1998) idea of 'cultures of landscape'. Here, 'landscape' circulates materially and symbolically in multifaceted cultural movements of identity, authority and belonging (Wylie 2007: 95). Throughout, I invoke the concept of a 'culturally charged' (Matless 2000: 142) landscape to engage with the tensions that emanate from Northern Ireland's – and, by extension, the mountain's – colonial past and ambiguous 'post-conflict' present. In this chapter, I explore Republican resistance to British militarism, as expressed at the crossroads of identity, culture and the shifting landscape of Divis Mountain.

This research project was prompted by my own lived experience of the mountain, as it underwent dramatic transformations in the years following the peace accords. I first visited Belfast on a travel fellowship in 1999, at which time I was accompanied to the summit of Divis by authorized personnel. Over the next decade, as the mountain was opened to the public, I became involved in efforts to make Divis accessible to Northern Ireland's increasingly diverse residents, primarily through leading walking tours for hard-to-reach community groups. At the same time, I became fascinated by the passion that local Republican residents expressed toward Divis and by the fierce sense of pride in 'their' mountain. The case study I present here focuses on the Republican heartland of West Belfast, where neighbourhoods that nestle at the base of Divis Mountain are intimately bound to the ongoing narratives of the Troubles. To explore these narrative voices, I adopted a methodological approach that combined archival research with semi-structured qualitative interviews. From August 2007 to August 2008, I conducted participant observation and 17 semi-structured interviews with a range of individuals, including former volunteers of the IRA, ex-prisoners, environmental activists, National Trust employees and volunteers, Irish-language specialists and artists. The interview participants – both male and female – comprised a broad age spectrum, ranging from young people in their mid-twenties to a group of elderly residents in a care home. Through their voices, I intended to glean from Republican West Belfast a range of experiences, memories and ways of relating to the cultural landscape of Divis.

This chapter explores the ways in which *ideas* of the mountain circulate in narratives of Republican identity that are emerging from the history of conflict, and are now evolving through the peace process. I am particularly interested in how the mountain serves as both a proxy for the imposition of British military identity, as well as the means through which Republican resistance is articulated. The first section traces the mountain's contested colonial history through two related dimensions of the British imperial project: the nineteenth-century

Ordnance Survey mapping process and the British army's controversial twentieth-century military presence. In the second section, I enlarge the imperial reading to explore the relationship between Divis Mountain and Republican resistance in the cultural heartland of West Belfast. The third section draws on the *Herald of Jericho*, a remarkable work of public art, to explore the complexities of memory, landscape and trauma in a place recovering from violent conflict.

Military geographies and imperial landscapes

As a post-colonial entity, Northern Ireland is an ambiguous case. In 1920, the Government of Ireland Act released from Britain the 26 counties that would eventually become the Republic of Ireland, thus signalling (at least in theory) a clear break in political colonial status. However, six counties in the northeast corner of the island remained within the United Kingdom. The creation of Northern Ireland, as it was now known, became a source of bitter conflict, the consequences of which would unfold over the course of the twentieth century.

Divis Mountain serves as a symbol of Northern Ireland's complex legacy of colonialism and its tensions around authority, memory, identity and change. These tensions can be explored through the mountain's fraught history of mapping, which in turn can be read in the wider context of geographical scholarship on imperialism, maps and power (Blunt and Wills 2000; Carter 1987; Harley 1988; Huggan 1989). The Ordnance Survey was one of the most extensive aspects of the British imperial project, with military surveyors dispatched to the far corners of the empire (see also Clayton 2008; Edney 1997; Seymour 1980). The mapping of Ireland began in July 1825 (Andrews 2002). Three years previously, surveyors had observed Divis Mountain from a vantage point in Scotland. Its summit became the first point for the interior triangulation of Ireland, which the surveyors combined with observations made from Scotland in order to frame Ireland's trigonometric skeleton. By linking Irish coordinates into the existing British triangulation system, the survey method sent a message that Ireland was part of Britain (Smith 2003: 84), thus implicating Divis Mountain in the strategic work of British imperialism.

Parallel to the material process of the Ordnance Survey, Britain continued to flex its imperial authority through the naming of places (Withers 2000), formalized through the Ordnance Survey's policy of anglicizing the spelling of place names (Andrews 1997: 300). Kiberd (1995: 619) describes the result as 'an English grid . . . remorselessly imposed on all Irish complexities'. At the edge of Belfast, itself an anglicization of *Beal Feirste* for 'mouth of the sandbars', the highest summit became known through the 'paper landscape' (Andrews 2002) as 'Divis'. The name is an anglicized version of *Dubh*, the Irish word for 'black', by which the mountain had been known to the local Gaelic-speaking population. Despite the map-makers' attempts to inscribe the landscape with British identity, local Republicans resisted by (re)claiming their original place names in daily use. To this day, many local residents insist on referring to the mountain as 'the Black Mountain' rather than 'Divis' – the name given on the Ordnance Survey map.

The role of Divis in the nineteenth-century imperial mapping process prefigured its military occupation during the twentieth century. During the Second World War, the British army maintained an active and controversial presence on Divis. Shortly after the war ended, the Ministry of Defence (MoD) formally leased a large swath of land, which included Divis, from local landowners. Later, during the height of the Troubles, it purchased this property outright.[1] On the summit of Divis, the MoD erected infrastructure for communications, including a series of aerial towers that could be seen from virtually anywhere in the city below. In nearby fields, soldiers honed their skills on newly created rifle ranges. And on the mountain's lower slopes, which once served as bleaching greens for the linen industry, a large army fort loomed over West Belfast. For local residents, the occupation of Divis by the British military became one of the most iconic injustices of the Troubles – not least because it provided the army with a superb base for exerting control over the Republican communities of West Belfast. One young man, born in the early 1980s, compared West Belfast during the Troubles to 'an open prison'. He described helicopters flying overhead, regular army patrols on the street and, crucially, the 'all-seeing eye' of Divis Mountain.[2]

In the late 1990s, as the Troubles were winding to a close, the MoD quietly began to examine its vast holdings. Divis Mountain became classified as 'surplus to requirements' (Green Balance 2006: 52). This admission was a startling reversal for what, in recent memory, was heavily militarized terrain. Not long after the re-classification, the MoD announced its intention to sell the property. Eventually, it was sold to a collection of public heritage bodies that, together, purchased the property with the intention of transforming Divis into public open space. On 24 June 2005, following several years of intensive cleaning and preparation, Divis Mountain was (re)opened to the public.

Tensions, however, continue to resonate in the mountain's complicated militarized legacy and its proxy role in the attempted imposition of British military identity. Even after the transfer of Divis to public ownership, the MoD retained nine acres of land on the summit – including the aerial towers built during the height of the Troubles. The MoD encircled their compound with a high fence topped with barbed wire. This fence, along with the highly visible aerial towers, served as a stark reminder of the continuing military presence on the mountain.

Only in 2009 – nearly five years after the original sale – did the summit convert fully to public use. Even then, despite the removal of the barbed wire fence, the imprint of the concrete base remains a visible reminder of the mountain's 'military geographies' (Woodward 2004). Divis may now belong to the public, but questions persist in the public imagination about the activities of British military personnel during their tenure on the mountain. These tensions amplify readings of Divis Mountain as a landscape of imperial control, through which – and *on* which – the British state exerted colonial rule in Ireland. Today, these tensions can be read in the contested landscape of Divis. The mountain continues to circulate in contemporary discourses of Republican memory and cultural identity – particularly those related to West Belfast's heritage of resistance. In the next section, I explore 'resistance' as a marker of Republican cultural identities that are deeply entwined with memories of Divis.

Narratives of Republican resistance

Identities of resistance offer rich analytical scope when explored through a geographical lens – in other words, in the context of place. I argue that Divis Mountain – as both physical presence and metaphorical touchstone – serves as a landscape through which the 'resistance communities' of West Belfast (Bean 2007: 57) articulate their cultural identities as Irish Republicans. As Pile (1993: 3) observes, identities of resistance are 'taken up not only in relation to authority . . . but also through experiences which are not so quickly labelled "power", such as desire and anger, capacity and ability, happiness and fear, dreaming and forgetting'. I argue that these emotional capacities, when grounded in the physicality of place, allow for more nuanced readings of resistance that complicate what Cresswell (2000) describes as the tendency toward romanticism. In this section, I explore some of the ways in which identities of Republican resistance are created, articulated and embodied in relation to Divis Mountain (Figure 8.1).

Over the past two decades, Northern Ireland's peace process has dramatically redefined contemporary Republicanism, from its origins in anti-colonial armed resistance to a movement that supports non-violent constitutional politics. On the ground, Republican ideology remains entrenched in both the idea and the material

Figure 8.1 Divis Mountain as viewed across West Belfast.

Source: Photo by Christoff Gillen.

fact of community, and Republican identity continues to find tangible expression in staunchly Republican areas. In West Belfast today, support for contemporary non-violent Republican values sits alongside devotion to the IRA. Its legacy is a potent source of pride and identity for Republican communities that endured the long civil war. The presence of former IRA volunteers, many of whom served prison sentences during the Troubles, serves as a living link to a glorified past.

Estimates for the number imprisoned are difficult to calculate, but some sources suggest approximate totals of 15,000 Republicans and somewhere between 5,000 to 10,000 Loyalists (McEvoy 2001, cited in Shirlow and McEvoy 2008: 2). Unlike their Loyalist counterparts, however, Republican prisoners captured the attention of the international media by continuing their campaign of resistance. In the notorious Maze prison (among others), they launched a raft of actions, aided by a sophisticated paramilitary command structure capable of reorganizing itself behind bars. The 'blanket protest' began in 1976, in which prisoners chose to wear their prison-issued blankets instead of the uniforms issued to criminal – rather than political prisoners (Coogan 1980, 1987; McEvoy et al. 2004). It escalated into the 'dirty protest', in which Republican prisoners refused to bathe and smeared the walls of their cells with their own excrement. Ultimately, the long-running 'blanket' and 'dirty' protests came to be regarded as self-defeating (McKeown 2001, cited in Shirlow and McEvoy 2008: 37) and were suspended by the IRA leadership. Nonetheless, the legacy of these protests contributes to the narratives of resistance that are still celebrated in Republican communities today.

For a former prisoner named Seamus,[3] memories of Divis Mountain are linked powerfully with those of the Maze prison. During the Troubles, Seamus spent over 11 years behind bars; his original sentence of 18 years was reduced to 9 years on remission, but prison authorities added time on to his sentence for each day in which he participated in the IRA's organized protests. Along with hundreds of his fellow prisoners, Seamus 'went on the blanket', participated in the 'dirty protest', and was present during the high-profile hunger strikes that led to the deaths of ten colleagues.

In his interview, Seamus spoke of a jarring juxtaposition between his oppressive memories of prison and those of Divis Mountain. In 1986 – ten years into his sentence – Seamus was escorted by prison authorities to a medical appointment in southwest Belfast. From the hospital's car park, Seamus saw the mountain for the first time in over a decade. He described to me how childhood memories of bluebells, gullies and sheep came flooding back to him. These vivid landscape memories contrasted sharply with the bleak interior world of the Maze, prompting Seamus to say to himself: 'I have to go back there . . . I want to go home.'[4] His unexpected encounter with Divis from the hospital car park, which he recounted over 20 years later, speaks to the emotional power of memory. Here, the mountain's sheer physicality concentrates his longing and becomes a symbolic point that focusses Seamus' memories of his past life. Heller (1995, cited in Morley 2000: 24) defines 'home' as 'awareness of a fixed point in space, a firm position from which we "proceed" . . . where we feel safe and where our emotional relationships are at their most intense'. For Seamus, Divis Mountain serves as this 'fixed point'. As he explains: 'You knew where it was . . . where you lived.'

The idea of 'home' recurred in many of the interviews I conducted – both in relation to Divis and to the larger idea of Ireland itself. According to Padraig, another former IRA combatant: 'For people my age, the mountain pre-dates the conflict.'[5] Although the British military dates its presence on Divis to the Second World War, it managed to co-exist peacefully for several years with local Republican neighbourhoods. Like Seamus, Padraig possesses vivid child-hood memories of Divis as an idyllic outdoor playground. But as Northern Ireland edged toward civil war, the British military presence intensified in response, and the nationalist neighbourhoods of West Belfast became a nexus for Republican resistance. Many of the children who had played on the mountain in calmer times now took up arms, while others resisted in subtler ways (see Dowler 1998). For example, a local environmental campaigner recounted to me how he would defy the British army by continuing to walk on the mountain. He remembers stumbling across soldiers participating in surveillance exercises, as they lay half-hidden in the bracken.[6]

Padraig grew up in the Turf Lodge estate, located at the edge of the mountain. Alongside his childhood memories of Divis are unpleasant associations with military violence. Padraig recalls that British soldiers would take local residents into the hills and 'give them a good hiding'. Moreover, memories of military surveillance are particularly potent. Padraig remembers cameras dug into the face of the mountain, and the chilling knowledge that from their vantage point at Fort Jericho, the army looked directly into his neighbourhood. In the following exchange, Padraig describes how memories of military surveillance entwine with those of his growing Republican activism:

LIA: What was it like, living at the foot of the mountain and knowing you were watched?

PADRAIG: As an active Republican in Turf Lodge, you had to learn how to move across the estate in a particular way.[7]

Padraig can trace his growing awareness of military surveillance alongside his emergence as an 'active Republican' – a euphemism for his participation in the IRA. He describes an intricate relationship between the mountain, with its British military observers, and his shifting experience with the local landscape. Like many other Republican activists, Padraig developed skills for hiding from the network of cameras that criss-crossed his estate. The mountain, as the vantage point of the enemy, now became dangerous to Padraig.

That the mountain is both a source of danger and a site of great affection speaks to a complex juxtaposition of identities. When I ask him to describe the relation-ship between Divis and his sense of identity, Padraig links the mountain to the development of his political consciousness: 'At the same time that I became an active Republican, the mountain became an escape.' He feels the same way toward it today: 'There are parts of the mountain that are as beautiful as Donegal.

And there's nowhere as beautiful as Donegal.' Although local residents could, with varying degrees of ease, gain access to some of the lower slopes of the mountain during the Troubles, it seems likely that the 'escape' to which Padraig refers is metaphorical in meaning. The word takes on greater significance considering that Padraig, like many of his compatriots in the IRA, spent several years in prison during the Troubles. Furthermore, he heightens the mountain's symbolic value by comparing it to Donegal – Ireland's westernmost county, which resides in the Republic. The partition of Ireland was the result of political gerrymandering, based on Unionist fears that the large Catholic populations in three of the nine northern counties (including Donegal) would complicate attempts to create a Protestant-dominated Northern Ireland (see Bardon 1992). The resulting political border effectively separated County Donegal from the rest of the Irish Republic, bounded to the east and south by three counties of Northern Ireland, and by the sea to the north and west. Donegal's isolation from the rest of the Republic, and the status of its Irish-speaking *Gaeltacht*,[8] evokes great affinity amongst Republicans living in Northern Ireland. Indeed, Donegal looms large in Padraig's imagination as a place of great beauty to which no other can compare.

For Padraig, Divis offers transformative personal properties. He describes 'conversations on the mountain that you wouldn't have anywhere else'. As an example, he recounts a walk he once took with a friend. On Divis, they spoke freely about poetry and about their lives with a depth that, once they had descended, they never experienced again. Intriguingly, Padraig then attempts to describe in another way these dimensions to himself that he discovered on the mountain: 'During the conflict, you would become a different person when you crossed the border to the 26 counties.' Here, Padraig describes how a new personal identity is created in the crossing of political borders, from British-ruled Northern Ireland to the 'homeland' of the Irish Republic – the '26 counties', as he calls it. He suggests that Divis replicates this sense of border-crossing, as though the mountain allows him to tap into dimensions of his identity only found outside of Northern Ireland. In this way, the mountain appears to connect him to wider possibilities of identity, linked to his Republican vision of a united, independent Ireland.

Republican ex-prisoners have embodied the emotional resonance of the mountain in a variety of ways. For example, after Divis was re-opened to the public, Padraig participated in a sponsored walk designed to raise funds for the controversial defence of former IRA colleagues. The 'Colombia Three' were arrested in Bogotá in August 2001 and accused by the Colombian government of training rebels from the Revolutionary Armed Forces of Colombia. In conceptualizing the event, Padraig and his fellow organizers recognized the mountain's power to garner attention: 'We came to the conclusion that people would remember a walk across the mountain much longer than [they would remember] a function.' And in 1998, as the peace agreements put to rest the need for armed conflict, a group of former IRA members and ex-political prisoners formed a social walking club called the 'Irish Ramblers Association'. Today, membership hovers between 30 to 50 people, and the Ramblers meet several times a year for walks on Divis. The founding of the club can be read in many ways: as a transformation of the IRA; as continued, symbolic resistance through its deliberate acronym; as reclamation of a mountain long controlled by the British army; and as all of these at once. For Cormac, one of the club's founding members, the British military occupation was a brief anomaly: 'The British were only here for a blink of an eye . . . We've held the mountain for thousands of years . . . We've had the mountain for longer than they had it.'[9] Cormac's use of the word 'held' is intriguing in its physicality, emphasizing both the mountain's physical constancy amidst the changing geo-politics of place and identity, as well as the importance of the acts of resistance that he and the Irish Ramblers embody in their contemporary presence on the mountain. The Troubles may have ended, but the mountain serves as a constant presence through which Cormac and his contemporaries continue to articulate their resistance to British imperial control.

As former IRA combatants and ex-prisoners, Padraig, Seamus and Cormac are obvious figures for examining Republican resistance. The IRA, which emerged in direct response to British colonial rule, is an excellent example of what Jacobs (1993: 217) calls a 'tectonic' form of resistance. For combatants, imprisonment was an emphatic, visible sign of their resistance identities, amplified by the protests that continued from behind bars. However, 'resistance' can also adopt subtler expressions that are 'articulated through a range of more fragmented formations in which stark oppositions give way to more complex subversions' (Jacobs 1993: 217). In contemporary times, as violence has abated in Northern Ireland, Republican resistance has adopted more nuanced dimensions, as war-time memories are reworked for a peacetime heritage. For Republicans of West Belfast, Divis Mountain has served as a site of both stark opposition and complex subversion – the latter sometimes finding expression in creative ways. In the next section, I explore resistance as an artistic process that forges material and metaphorical links between the transforming landscape and contemporary memory-making.

The Fort and the *Herald*

One remarkable expression of Divis in Republican cultural identity can be found in a public art sculpture created in 2002. *Herald of Jericho* (Figure 8.2) hangs in the foyer of the Upper Springfield Development Trust, a community regeneration and economic development charity based at the westernmost edge of Belfast.

The organization's remit encompasses the densely populated neighbourhoods that once stretched across the linen industry's vast, rural bleaching greens. Deirdre Mackel, Arts Officer for the Trust, coordinates public art projects for West Belfast, whose residents tend to face economic and social barriers to visiting the galleries located in the city's more affluent areas. Divis Mountain is of particular interest to Deirdre, who frequently incorporates its image and story into her community projects, of which the *Herald of Jericho* is a dramatic example. For Frank Quigley, the artist commissioned to lead the project, the process of conceptualizing and producing the sculpture is intimately entwined with his activist Republican identity. During the Troubles, he too served time in the Maze prison for his

Figure 8.2 Herald of Jericho.

Source: Photo by Lawrence Kirk.

association with the IRA. The sculpture's story of emergence – its provenance, conceptualization, production and reception – illuminates the links between culture, identity and heritage in the (de)militarization of Divis Mountain.

The *Herald*'s origins lie in the military forts of the Troubles. By the 1970s, the British army occupied three bases in West Belfast, known as Forts Monagh, Pegasus and Henry Taggart. According to Padraig (introduced in the previous section), the three bases were all 'within eyesight of each other'. To the Republican communities of West Belfast, these military bases were sources of resentment that proclaimed British military power in symbolic and material form. For example, the army built Fort Pegasus on the grounds of the Gaelic Athletic Association, a site of deep cultural significance to local Republican communities. Symbolically, and along similar lines to the nineteenth-century Ordnance Survey, these three forts can be read as attempts to assert British military identity on the local landscape.

Despite the proximity of their existing army bases in this corner of Belfast, the British military decided to spearhead yet another. Fort Whiterock, located at the eastern edge of Divis Mountain, was the last and largest operational base built in a nationalist area of Belfast. To amplify the controversy, the land identified for the project was an industrial cooperative of workshops owned by local residents. In the face of soaring unemployment, the Whiterock Industrial Estate had embodied the economic hopes of residents in West Belfast. Consequently, the army's decision to claim this site provoked further outrage amongst local Republicans. From its position, high in the hills and backed by the mountain, the fort offered superior surveillance opportunities for observing Republican activists in the neighbourhoods below. In November 1979, the British army moved on to the site. During its construction, the soldiers had erected temporary metal walls to hide from view the layout and building progress. The wind, however, kept blowing down these walls, evoking the ancient city of Jericho and the biblical story of its fall. To the local residents, Fort Whiterock became widely known as Fort Jericho (see de Baroid 2000). Padraig described his development as a Republican activist alongside the erection of Fort Whiterock/Jericho. When construction began in the late 1970s, Padraig was a fifteen-year-old student living in the nearby Republican stronghold of Turf Lodge. He took an active role in protesting the fort's presence, attacking it every day on his way to school. Once built, Fort Jericho blocked the view of the mountain for decades. According to Frank Quigley: 'It was a huge, grey lump . . . Just a blot on the landscape, to use a cliché. The mountain just disappeared. You wouldn't have seen the mountain after that.'[10]

For today's younger generation – those who were children in the 1980s and 1990s – the military bases formed their known landscape and created a more complicated dynamic of resistance than that embraced by their elders. For example, I interviewed twenty-six-year-old Aidan about his experience of growing up near Fort Jericho in the 1990s. Below, he describes a nuanced relationship between the soldiers and local teenagers:

> And we would've sat there, right below them. There was a patch where we would've played football . . . There was a couple of British soldiers who were friendly . . . Now, a lot of us had a policy of not talking to them. We just ignored them . . . Some of the kids that I can remember that we played with, they weren't in any way political. They just took things as they were and they did speak to them . . . There were young soldiers who supported Celtic,[11] and they would've thrown fags [to us], and some of the kids would've ended up giving them a cigarette back and so on.

In this excerpt, Aidan illustrates that amongst his contemporaries there was a diverse range of opinions about, and ways of engaging (or consciously *dis*engaging) with, the British military presence. Aidan acknowledges that this diversity may be due to a generational shift: 'I imagine in the 80s it wouldn't have happened. There would've been less civil relationships.' Yet beneath these friendly interactions there simmered another dynamic, which erupted in the tense period leading to the 1994 ceasefire declaration:

> So we would've been throwing petrol bombs at [the soldiers] rather than throwing cigarettes at them . . . I can remember when the IRA ceasefire was called, the 31st of August '94 . . . There was a statement put out [that] from midnight, there would be a cessation of all the military operations. But about 10 o'clock, we were all going down to the fort, and the IRA came and attacked the fort with coffee jar bombs and stuff like that . . . The next day, kids gathered around to see the damage . . . That was great, come in the next day and see what damage there is.

As the friendly exchange of cigarettes gave way to destruction by petrol bombs, Aidan and his friends were drawn into the militant political struggle of their elders. Although Aidan is too young to remember a time before Fort Jericho, the enthusiasm that he recalls when recounting the damage speaks to the strength of narratives injustice in Republican culture, and the memories that cross generational lines.

On 18 May 1999 – after nearly 20 years of military occupation – the British army drove the last of its vehicles through the gates of Fort Jericho, returning what remained of the Whiterock Industrial Estate to the local Republican community (de Baroid 2000: 386). Demolition began shortly thereafter, transforming the physical landscape and offering residents of West Belfast their first unobstructed view of the hills since the fort's inception. As Frank Quigley remembers, '[i]t just went from grey steel to green mountain when they were finished'. His observation was echoed in the workshops that Deirdre Mackel organized for local residents, in which she asked them to conceptualize the sculpture that would eventually emerge as *Herald of Jericho*. According to Deirdre, 'people that we were talking to said: "Well, [we've] got the view of Black Mountain back".'[12] As I discussed in the first section, many local residents have resisted the Ordnance Survey's imposition of 'Divis', choosing to refer instead to the 'Black Mountain'. Deirdre's comment speaks to the strength of the semantic resistance practised by the Republicans who lived – and continue to live – near the mountain.

The demolition of Fort Jericho represented a long-delayed triumph for Republican West Belfast. Frank recalls: 'It was a certain victory. The British army hadn't won. They left before the IRA did.' Nearby, from her office at the newly built premises of the Upper Springfield Development Trust, Deirdre conceived an idea for creating a public art sculpture from materials sourced from the fort's ruins. With difficulty, she arranged for clearance to enter the demolition site and to gather materials. Deirdre recalls the control that the MoD continued to exert, even as it dismantled its stronghold: 'We had to go in and just tell them which bits we wanted.'

To design the sculpture, Frank and an Irish-Canadian artist named Farhad O'Neill consulted with groups of ex-prisoners and young people. Through talking with one another and through sharing their memories, the idea for *Herald of Jericho* took shape. Deirdre describes the concept as one of transformation and reclamation:

> The notion was taking something from a bad period in history and doing something positive with it . . . and it was significant for people of the area that this fort was being dismantled. It was a way to say, 'This is a new time now. Good-bye' . . . This provided a symbolic way of doing that . . . Heralding the new dawn.

Frank and Farhad (a metal-working specialist) proceeded to collaborate with local residents to create the sculpture. From the materials gathered from the demolished army base – palisades, corrugated metal, rocket wire – the artists and residents cut, welded and fashioned a large metal angel with outstretched wings, holding a bugle in her hand. They inscribed the angel with Republican identity, through decorative Celtic designs and, less visibly, through inscriptions of their own names. Twelve ex-prisoners, many of whom Frank knew from his own time at the Maze, lifted the sculpture into place, where it now hangs from the high-ceilinged atrium of the Upper Springfield Development Trust.

In February 2002, *Herald of Jericho* took flight. Those present at its launch included local residents, ex-prisoners, former members of the industrial cooperative, representatives of the Arts Council of Northern Ireland, and Gerry Adams, a well-known Republican activist and former prisoner, who went on to serve as a member of Northern Ireland's power-sharing government. Also present was Father Des Wilson, a priest whose active presence in West Belfast during the Troubles, against the wishes of the Irish Catholic Church, is now legendary (de Baroid 2000). At the launch, Father Wilson's speech encapsulated the narrative of transformation embodied in the sculpture:

> There is a great saying in our tradition, that swords will be melted down and made into ploughs, that the weapons of war will be made into instruments of prosperity, and here we have a piece of an army barracks reworked into a symbol of hope and work and prosperity.
>
> (*Irish News* 2002)

Yet this 'symbol of hope and work and prosperity' is a bittersweet and incomplete memorial to recent history. When sunlight filters through the foyer, the angel's skirt of palisades casts shadows shaped like prison bars, evoking the memories of place and the embodied experiences of the prisoners who helped to design and position the sculpture.

The demolition of Fort Jericho can be interpreted as a microcosm of the larger demilitarization of Divis. These transformations, although widely celebrated by local residents, nonetheless trail darker memories in their wake. Deirdre reads these shadows as a narrative for the larger Republican community:

Well, the fort, when it was there, it was so imposing . . . It obliterated the view of the Black Mountain, but also cast shadows that aren't seen on the people. Like a negative impact on people's lives. So whatever way [the sculpture] is lit, it casts shadows.

The *Herald of Jericho* captures some of these complex, emotional resonances. Its creation from the fallen walls of Fort Jericho, reworked by and for the community, underscores continued resistance to the identities of British military presence in West Belfast. The sculpture's story illuminates the complex efforts to reclaim the mountain for and by contemporary Irish Republican culture, and to inscribe West Belfast as a place where resistance continues to be practiced in myriad, creative forms.

Conclusion

Today, the young men and women who came of age during the Troubles are middle-aged leaders of their communities. Their defiance continues to energize West Belfast, and their memories of resistance shape the contemporary cultural landscape. This striking heritage finds expression through murals, memorials and, for many, the iconic presence of Divis Mountain. As the narratives of this chapter suggest, Divis serves as a fixed physical point amidst many shifts of interpretation: the mountain as a site of Gaelic heritage; as powerhouse for historic industry; as proxy for the imposition of British militarism; as the vantage point of 'the enemy'; as both site and resource for resistance; and (most recently) as public parkland. In its metaphorical malleability, Divis can also be read as what Morley (2000: 44) describes as a 'synecdoche for the unreachable lost home.' Indeed, the theme of home and homecoming resonates throughout the narratives illuminated here.

As a cultural icon, Divis raises interesting questions about the relationship between resistance and landscape. As the memories excavated here suggest, resistance is as much about the reclamation and reinscription of place as it is a series of actions. These narratives – powerfully encapsulated in the *Herald of Jericho* – reach in multiple temporal directions at once, backwards and forwards in time. As Morin (2005: 319) points out: 'Debates about landscape do not simply rest on what landscape *is* but also on what landscape *does* – how it is produced and how it works in social practice' (original emphasis). In other words, landscape is anything but static: it is an ideological and symbolic process that holds the power to reproduce cultural and social identities. As Northern Ireland recovers from decades of violent conflict, Republican communities of West Belfast remember and recreate Divis as a place of resistance – one where a familiar yet ever-shifting landscape is continually reinscribed with identities of *home*.

Notes

1 Interview, National Trust, 2 June 2008.
2 Interview, 6 November 2007.

3 All names of respondents in this section have been changed.
4 Interview, 5 June 2008.
5 Interview, 27 June 2008.
6 Interview, 2 November 2007.
7 Interview, 27 June 2008.
8 *Gaeltacht* refers to an Irish-speaking geographical entity. In Ireland, the *Gaeltacht* refers to the regions, located primarily in the west of the island, where the Irish language is still spoken.
9 Interview, 19 October 2007.
10 Interview, 4 July 2008.
11 Celtic Football Club is a football team located in Glasgow, popular with Republican communities in Northern Ireland.
12 Interview, 13 August 2008.

References

Andrews, J.H. 1997. *Shapes of Ireland: Maps and their Makers (1564–1839)*. Dublin: Geography Publications.

Andrews, J.H. 2002. *A Paper Landscape: The Ordnance Survey in 19th Century Ireland*. Dublin: Four Courts Press.

Bardon, J. 1992. *A History of Ulster*. Dundonald, Northern Ireland: Blackstaff.

Bean, K. 2007. *The New Politics of Sinn Féin*. Liverpool: Liverpool University Press.

Blunt, A. and Wills, J. 2000. *Dissident Geographies: An Introduction to Radical Ideas and Practice*. London: Prentice Hall.

Carter, P. 1987. *The Road to Botany Bay*. London: Faber and Faber.

Clayton, D. 2008. Imperial geographies. In J.S. Duncan, N.C. Johnson and R.H. Schein (eds) *A Companion to Cultural Geography*. Oxford: Blackwell, 449–468.

Coogan, T.P. 1980. *On the Blanket: The H Block Story*. Dublin: Ward River Press.

Coogan, T.P. 1987. *Disillusioned Decades*. Dublin: Ward River Press.

Crang, M. 1998. *Cultural Geography*. London and New York: Routledge.

Cresswell, T. 2000. Falling down: resistance as diagnostic. In J.P. Sharp, P. Routledge, C. Philo and R. Paddison (eds) *Entanglements of Power: Geographies of Domination/ Resistance*. London: Routledge, 256–268.

de Baroid, C. 2000. *Ballymurphy and the Irish War*. London: Pluto Press.

Dowler, L. 1998. 'And they think I'm just a nice old lady': women and war in Belfast, Northern Ireland. *Gender, Place and Culture* 5 (2): 159–176.

Edney, M. 1997. *Mapping an Empire: The Geographical Construction of British India, 1765–1843*. Chicago: Chicago University Press.

Green Balance. 2006. *The Disposal of Heritage Assets by Public Bodies: A Report by Green Balance for The National Trust*. London: The National Trust.

Hardesty, D.L. 2000. Ethnographic landscapes: transforming nature into culture. In A.R. Alanen and R.Z. Melnick (eds) *Preserving Cultural Landscapes in America*. Baltimore and London: Johns Hopkins University Press, 169–185.

Harley, J.B. 1988. Maps, knowledge, and power. In D. Cosgrove and S. Daniels (eds) *The Iconography of Landscape*. Cambridge: Cambridge University Press, 277–312.

Heller, A. 1995. Where we are at home. *Thesis Eleven* 41.

Huggan, G. 1989. Decolonizing the map: post-colonialism, post-structuralism and the cartographic connection. *Ariel* 20: 115–131.

Irish News. 2002. Flying angel heralds the end to Jericho's walls. 26 February, p. 3.

Jacobs, J.M. 1993. Resisting reconciliation: the secret geographies of (post)colonial Australia. In S. Pile and M. Keith (eds) *Geographies of Resistance*. London: Routledge, 203–218.

Kiberd, D. 1995. *Inventing Ireland: The Literature of the Modern Nation*. London: Vintage.

McEvoy, K. 2001. *Paramilitary Imprisonment in Northern Ireland: Resistance, Management and Release*. Oxford: Oxford University Press.

McEvoy, K., Shirlow, P. and McElrath, K. 2004. Resistance, transition and exclusion: politically motivated ex-prisoners and conflict transformation in Northern Ireland. *Terrorism and Political Violence* 16 (3): 646–670.

McKeown, L. 2001. *Out of Time: Republican Prisoners 1970–2000*. Belfast: Beyond the Pale Press.

Matless, D. 1998. *Landscape and Englishness*. London: Reaktion.

Matless, D. 2000. Action and noise of a hundred years: the making of a nature region. *Body and Society* 6 (3–4): 141–165.

Morin, K.M. 2005. Landscape and environment: representing and interpreting the world. In S.L. Holloway, S.P. Rice and G. Valentine (eds) *Key Concepts in Geography*. London: Sage, 319–334.

Morley, D. 2000. *Home Territories: Media, Mobility and Identity*. London: Routledge.

Pile, S. 1993. Introduction: opposition, political identities and spaces of resistance. In S. Pile and M. Keith (eds) *Geographies of Resistance*. London: Routledge, 1–32.

Seymour, W.A. 1980. *A History of the Ordnance Survey*. Folkestone: Dawson.

Shirlow, P. and McEvoy, K. 2008. *Beyond the Wire: Former Prisoners and Conflict Transformation in Northern Ireland*. London: Pluto.

Smith, A. 2003. Landscape representation: place and identity in nineteenth-century Ordnance Survey maps of Ireland. In P.J. Stewart and A. Strathern (eds) *Landscape, Memory and History: Anthropological Perspectives*. London: Pluto, 71–88.

Withers, C.W.J. 2000. Authorizing landscape: 'Authority,' naming and the Ordnance Survey's mapping of the Scottish Highlands in the nineteenth century. *Journal of Historical Geography* 26 (4): 532–554.

Woodward, R. 2004. *Military Geographies*. Oxford: Blackwell.

Wylie, J. 2007. *Landscape*. London: Routledge.

9   In the shadow of centenaries

Irish artists go to war, 1914–1918

Nuala C. Johnson

As we are in the throes of a decade of centenaries across Britain and Ireland remembering the hundredth anniversary of the First World War, as well the Easter Rising in 1916, the subsequent war of independence and partition of Ireland under the terms of the Government of Ireland Act 1920 and the Anglo-Irish Treaty signed in 1921, the commemorative impulse is being enacted through a huge variety of memory-making practices and events (History Ireland 2014). From fresh academic studies to popular and official acts of remembrance, this period has stimulated numerous re-examinations of the broader social, political and cultural impact of conflict in shaping European and extra-European identities, territories and geopolitical relationships. A new library of academic narratives is emerging providing enlivened insights into the causes of the war and the actions of different combatant states; a host of television documentaries on these events has been commissioned and broadcast; vast numbers of community-led peoples' history projects are being undertaken; and new museum exhibitions, dramas, movies and literary interpretations of the period are emerging (Evans 2014). Of course the primary difference between now and earlier acts of commemoration is that the Great War is now outside of living memory and our connections to it are more indirect, mediated and diverse. In Ireland [both north and south of the border] efforts to mark the centenary are intense and have become part of the narrative of reconciliation, inclusion and re-absorption of this period of history into popular memory especially in Northern Ireland (Grayson 2010). One area of investigation that has received limited scholarly attention is the critical response of the visual arts to the war, and, in particular, the work of Irish-born painters who served as official war artists during the course of the conflict (Jeffrey 2014). Yet they have left us with a powerful visual archive of the war from both the perspective of the battlefront and the home front, and form part of an iconic record of these events as they happened in the second decade of the twentieth century. Moreover they also represent a coterie of artists whose relationship with Ireland was mediated through a complex set of alliances, allegiances, antipathies and identities in the liminal space that the island represented during that decade.

This chapter will consider two of the most significant Irish war artists, the Dublin-born Protestant William Orpen and the Belfast-born Catholic John Lavery. I will examine some of their war paintings and situate them in the context of a transforming set of Anglo-Irish relations that would characterize the decade from the First World War's beginning through to Irish independence. Combining a discussion of their biographical histories with their engagement with the Great War, in this chapter I will highlight the significance of the geographies of allegiance in conjugating personal and political identities. The First World War marked an important defining moment in relations between Ireland and Britain, not least in terms of the role of the Easter Rising in 1916 and the subsequent execution of its leaders, in situating the war in the Irish popular imagination and post-war memory (Jeffrey 2000; Johnson 1999, 2007; Pennell 2012). Moreover war art would become one of the principal ways in which visual memories of the war would be translated to popular audiences. While photography and film were to play some role in providing an archive of imagery upon which the wider public would remember the war, it was painting that endured as the powerful medium for communicating descriptively, as well as symbolically and ethically, responses to the war effort and connecting it with the cultural-political identities of the different combatant 'nations'. In the case of Orpen and Lavery, the channelling of their creative talent across a range of artistic subjects will illuminate how they related to the European conflict and the conflict in Ireland. While their war art had no particularly Irish content per se, in contrast to the work, for instance, of the Belfast-born William Conor, events in their homeland did not go unnoticed (Jeffrey 2014). More broadly, by focusing on the manner in which the arts have engaged with the war we can begin to echo the Irish poet Seamus Heaney's claim that, 'I can't think of any case where poems changed the world, but what they do is they change people's understanding of what's going on in the world' (Heaney 2004: 3). It is the cognitive, affective and emotional registers that the arts in general and the visual arts in particular have to offer, in providing an interpretive apparatus for reacting to conflict, that makes them both enduring and relevant in analysing discourses of memory. While war memorials and the rituals of remembrance surrounding public commemorative activities have received widespread scholarly focus in drawing attention to the emotional geographies of war (Gordon and Osbourne 2004; Johnson 2002; Kidd and Murdock 2004; Winter 1995), war art has received rather less sustained research. But as Anderson (2013: 456) has reminded us in his recent overview of the literature on the geographies of affect and emotion, we need to start thinking about 'how images work performatively as devices that move bodies affectively'. This chapter seeks to begin this process by examining some key artistic works produced during the war that brought it pictorially and discursively into the lives of those, particularly on the home front, affected by the war, and entered into the long-term collective memories of British and Irish society in the calibration of ideas of nationhood and national identities.

Enlisting the arts

Shortly after the war's outbreak, the Chancellor of the Exchequer, David Lloyd George, was tasked with establishing the British War Propaganda Bureau, better known as Wellington House, the building in which it was headquartered in central London. Responding to the knowledge that Germany already had established a propaganda agency, the imperative to channel Britain's propaganda effort through an official department was realized and Liberal MP Charles Masterman headed up the bureau (Sanders 1975). Initially Wellington House devoted its time, in secrecy, to publishing and distributing pamphlets and books that promoted the government's view of the war. Enlisting some of the most significant writers of the day to support and work with the agency, including people such as Arthur Conan Doyle and Henry Newbolt, it produced over 1,100 pamphlets during the course of the war. These pamphlets supported Britain's role in the conflict and were directed at neutral states around the globe, particularly the United States (Taylor 1999). A pictorial section was established in the bureau in 1916 in order to visually capture some arresting imagery of the conflict as experienced at the front and at home. This visual propaganda included the production of lanternslides, postcards, calendars, photographs, bookmarks and line drawings for worldwide distribution. The appetite for images, especially of the sites of battle, was intense especially among the editors of newspapers and illustrated publications, as well as for the department itself. Photographs, however, were not meeting this huge demand for a pictorial record of the conflict. By 1916 newspapers were exploring ways to procure new images for use in their publications and were offering monetary rewards to soldiers on the front to supply suitable line drawings. Moreover Masterman had learned in the summer of the same year that the renowned Scottish etcher, Muirhead Bone, had been called up for military service. He thought that the services of Bone might be better utilized through using his artistic talent to produce images for Wellington House than as an infantry officer. Consequently Bone was recruited as the first official war artist. He was sent to France and by October had produced over 150 drawings of life on the front. On his return he was replaced by his brother-in-law, the portrait painter Francis Dodd, who completed more than 30 portraits of senior military figures.

Bone's drawings were published in ten monthly parts, beginning in 1916, and they proved very popular with the public as they were both affordable and accompanied by an explanatory essay (Malvern 2004). Collecting war memorabilia of a visual nature had begun. Moreover they confirmed Masterman's view that they would serve 'as a novel adjunct to the programme of pictorial propaganda' (Gough 2010: 23). As France and Germany had already recruited war artists to the battlefront, the appointment of Muirhead Bone as Britain's first official war artist meant the government was following a wider trend among combatant states to maintain a visual record of the conflict (Hynes 1990). War art would also provide a permanent and long-standing visual evocation of the Great War in the imagination and memory of the public in the post-war period.

The move to establish an official war-artist movement was also spurred on by the critical and popular acclaim that artist-soldier Eric Kennington's 'The Kensingtons at Laventie' (a reverse painting on glass) received when he exhibited it in London in April 1916 (Weight 1986). *The Times* reviewer captured the essence of the painting's impact: 'The picture convinces us that it is real life, but it is not at all like a photograph of the actual scene' (*The Times* 1916: 4). Such adjudication indicated the potential for art to provide a visual record of the war that not only engaged the viewer with the material dimensions of war but also, significantly, with the emotional and moral landscape of the conflict. Painting was seen to have the potential to generate an affective response to the battlefront that could be more compelling and transparent than photographic reproduction. As such, the capacity of art to elicit popular support for the war was recognized at the official level. The decision to establish a new Department of Information, responsible for propaganda, and headed by the diplomat, historian and novelist, John Buchan, in February 1917, and drawing on the expertise of key figures in the London arts scene, reinforced the state's commitment to using the visual arts to represent the war and to re-deploy many artists already in the armed forces to serve their country as war artists.

While the initial intention of the scheme was to create a pictorial record, the appointment of Lord Beaverbrook, William Maxwell Aiken, as Director of the new Ministry of Information in March 1918 (which replaced the Department of Information), brought the longer-term commemorative role of war art to the fore. Beaverbrook, with his business and newspaper experience, aimed to make the artistic representation of the war serve a wider purpose as a memorial legacy to the conflict for future generations to appreciate. Beaverbrook brought this ambition with him from his experience of leading the Canadian War Memorials scheme. He sought to guide the Ministry towards a wider remit of deploying the visual arts as an act of commemoration, the performance of which would have long-lasting effects on the public's imagination (Cooke and Jenkins 2001; Harries and Harries 1983). War art would become an archive as well as a propaganda tool and thus become part and parcel of a wider memorializing effort. As Gough (2010: 31) explains: 'Arguably the greatest legacy of the war's art was the scheme itself: under Beaverbrook's tutelage the Ministry for Information protected and promoted emerging artists, brought intellectual coherence to a previously haphazard programme of commissioning.' It was within this political and cultural context that many of the most significant names in British war art were to emerge. From John and Paul Nash to Stanley Spencer, artists produced a large collection of exceptionally alluring and, at times, harrowing images of life on the battlefront. These works went on to become some of the most iconic visual images of the First World War and their longevity is confirmed by their recirculation and reproduction in books, films, websites and galleries devoted to interpreting the material and cultural context of the war and its afterlife. In particular, but not exclusively, this archive is housed at the Imperial War Museum in London (Kavanagh 1988; Malvern 2000; www.iwm.org.uk/collections-research/about/art-design). It is also within this official context that two Irish-born artists would make their contribution to the visual memory of the conflict but who would also be torn, to varying degrees, emotionally and ideologically, by the turbulent politics of Ireland in the second decade of the twentieth century. It is to the contribution of William Orpen and John Lavery to the canon of war art as memorial devices that I now wish to turn.

Capturing the battlefront: William Orpen, 1878–1931

The Dubliner William Orpen was born in 1878 into an affluent, Protestant family and lived his early years at the family home, Oriel House in Stillorgan, a well-to-do suburb of the city. Although his father was a successful solicitor, both his parents and his eldest brother Richard were accomplished amateur artists and Orpen's drawing talent was quickly recognized and nurtured by his parents. At the age of 12 he enrolled in the Dublin Metropolitan School of Art, and, although the art school specialized in industrial design, his painting skills were quickly recognized and he won many local awards. In 1897 he moved to the Slade School in London (1897–1899) where he flourished and met many of the young promising artists of the day, including Augustus John. It was during his years at the Slade that Orpen developed his skills in portraiture, which 'may have derived from [t]his early impulse to commemorate and comment in paint upon those around him' (Upstone 2005: 10). He exhibited his early oil paintings at the New English Art Club in 1900 and they won much critical acclaim. While of small physical stature [approximately 5 feet 2 inches tall] and harbouring a negative self-image of his appearance, Orpen completed numerous self-portraits during his lifetime, including one of himself preparing for work at the Western Front. He married Grace Knewstub, the daughter of a London art dealer, in 1901 but he would have numerous affairs for the remainder of his life both with society ladies whom he painted and the models he used in his studio.

As his reputation was expanding in Britain, Orpen spent time teaching at the Dublin Metropolitan School of Art (between 1902 and 1914), at yearly or bi-yearly sessions. This brought him back to the home of his birth and into dialogue with friends and students with political allegiances different from his own. He was eager to transform the School from one specializing in industrial design to one focused on fine art. During these years, Orpen encountered students of a nationalist persuasion, in particular Sean Keating, who became, for a time, his studio assistant in London and who discouraged Orpen from any involvement in the Great War (Turpin 1979). At the school he also became friends with a gifted Protestant art student Grace Gifford, whom he captured in the portrait entitled 'Young Ireland'. He encouraged her to attend the Slade School, which she did from 1907–1908, but Gifford returned to Ireland after her training and supported the increasingly influential nationalist movement. She joined Sinn Féin, converted to Catholicism, and became an active promoter of the Irish language and the cultural revival. She got engaged to Joseph Plunkett, one of the leaders of the Easter Rising in 1916, and married him in his prison cell on the eve of his execution that year (Upstone 2005). Nonetheless she maintained her friendship with Orpen, now domiciled in England, who would have been well aware of her political commitments. Although Foster claims that 'Protestant families like the Orpens lived at a distance from their Catholic neighbours, even those of the same class' (Foster 2005b: 63), it is nonetheless clear that well-to-do Protestants did come into close professional and personal contact with Catholics and with those who did not necessarily share their social or political status or views.

Figure 9.1 William Orpen, 'The Holy Well', 1915.

Source: Photograph courtesy of the National Gallery of Ireland.

Orpen became a close friend of the collector and art dealer Hugh Lane and shared his ambition to create a museum of modern art in Ireland. At the same time he supported the labour leader Jim Larkin's campaign to improve the employment conditions for the Irish working class. Having said that, many of his paintings with Irish subjects reflected a critique of Irish piety, puritanism, the influence of the Catholic Church and the stifling bureaucratic structures inhibiting, in his view, progress towards modernity. His ambivalence towards the Ireland of his day is clearly expressed in his 1915 painting 'The Holy Well' (Figure 9.1).

In the composition his friend Sean Keating is modelled as the quintessential Connemara man from the west of Ireland, overlooking, with an air of disgust, the naked, head-bowed figures below him approaching the well of absolution overseen by a priest. Similarly in the 1913 canvas 'Sewing New Seed', Orpen's attitude toward his homeland is expressed. The painting is conceived as an allegory of the stultifying effects of an Irish bureaucracy resistant to change or innovation in the arts in Ireland. Thus while Orpen's pre-war engagement with Ireland was complex and ambivalent, it is perhaps noteworthy that after 1915 he never visited the island again apart from one day-visit in 1918. Foster suggests that Orpen represented that class of Protestants who felt completely Irish while adhering to the union with Britain, but where 'the "old Ireland" that accommodated the easy sense of belonging in both countries (and sustaining a privileged position in both), which Orpen grew up with, was long gone' (Foster 2005a: 40). Of course Britain too was undergoing radical social and cultural upheaval on the eve of the war, and the conflict itself would presage both short- and long-term transformative effects on the political and ideological landscape of the country (Eksteins 2000; Fussell 2000; Higonnet et al. 1989; Winter 2003).

By the war's outbreak, Orpen was one of the most acclaimed society portrait painters in Britain but Upstone claims that '[t]he Great War marked a watershed in Orpen's life; he was never the same after it' (Upstone 2005: 34). As a celebrated and popular artist, with significant connections to the British establishment, Orpen enlisted for the Army Service Corps in December 1915, where he was commissioned as Second Lieutenant and stationed at the Adjutant's Office in Kensington Barracks. Sean Keating had quit Britain prior to conscription and advised Orpen to '[c]ome back with me to Ireland. This war may never end . . . I am going to Aran. There is endless painting to be done' (Keating 1937: n.p.). With the introduction of the war artists' scheme, and through his personal contacts with senior statesmen, Orpen quickly secured himself a position as an official army painter and was posted to France. During this time he produced one of the largest corpus of paintings of any of the official war artists and subsequently donated all his war work to the state.

In April 1917 Orpen arrived in France and stayed until March 1918. Unlike other war artists who were permitted three weeks in France to prepare their work, Orpen was set no time limitations and, unusually for a war artist, was promoted to the rank of Major. He was provided with a car and a chauffeur by the Department of Information and he self-funded the services of a batman and a private secretary. These special privileges reflected his position as one of the most respected living artists but also put him under pressure to produce a body of work that reflected this status. He was mainly stationed in towns behind the front lines although most of his early paintings consisted of portraits of generals, senior staff officers and celebrated airmen from the Air Corps. This first phase of his work in France included portraits of Sir Douglas Haig and General Trenchard, mirroring his commercial success as a portrait painter and providing material that appealed to those in charge of the visual propaganda machine. As well as painting in France he maintained a lively correspondence with family and friends at home and recorded his experiences in a memoir entitled *An Onlooker in France 1917–1919*. His first impressions of the Somme, arriving there only three weeks after the Germans had retreated behind the Hindenburg Line, focused on the physical destruction of the landscape:

> I shall never forget my first sight of the Somme battlefields. It was snowing fast, but the ground was not covered, and there was this endless waste of mud, holes and water. Nothing but mud, water, crosses and broken Tanks; miles and miles of it, horrible and terrible.
>
> (Orpen 1921: 16)

Orpen's stationing in France coincided with the Department of Information's discussion about what war artists were to paint in the field. First and foremost it was declared that war artists should document and record the war. As well as producing portraits of senior military personnel during the early months of his arrival in France, Orpen also produced some line sketches and watercolours of individual soldiers on the front line during April and May 1917. 'A Man in a Trench' (1917) represents one of these moving studies of a young soldier facing his potential mortality on the battlefield, and reflects a wider societal concern that the war was destroying a whole generation of youths. While other war artists were producing more interpretive representations of the front through recon-structed landscapes of destruction, and with soldiers 'going over the top', Arnold (1981: 322–323) observes that 'Orpen concentrated on the direct and factual encounter'. When he returned to the Somme in the late summer of 1917, he was struck by how much the landscape had been radically transformed since his April experience. He recounted:

> Never shall I forget my first sight of the Somme in summer-time. I had left it mud, nothing but water, shell-holes, and mud—the most gloomy, dreary abomination of desolation the mind could imagine; and now in the summer of 1917, no words could express the beauty of it. The dreary, dismal mud was baked white and pure—dazzling white. White daisies, red poppies and a blue flower, great masses of them, stretched for miles and miles. The sky a pure dark blue, and the whole air, up to a height of about forty feet, thick with white butterflies; your clothes were covered with butterflies. It was like an enchanted land; but in the face of fairies there were thousands of little white crosses, marked 'Unknown British Soldier' for the most part.
>
> (Orpen 1921: 31)

This transmogrified landscape drove Orpen to transform his approach to representing the battle zone. He altered his palette of colours to include mainly pastel shades: white, pea greens, soft lavenders and mauves, and clear blues for

Figure 9.2 William Orpen, 'Dead Germans in a Trench', 1918.

Source: © Imperial War Museum (Art.IWM ART 2955).

depicting the skies. This change of mood in his work enabled him to express pictorially nature's capacity to rejuvenate amid a landscape of human destruction. As Gough (2010: 173) observes,

> an unusually piercing, acute and intense light that deflected off the seared white chalk casting bizarre shadows and extreme shifts in tone and colour. To

an artist accustomed to the dusky opulence of cavernous Edwardian drawing rooms it came as a revelation.

Dispensing with half tones and deploying shadowing and foreshortening to create effect, Orpen evoked this summer scene. 'Dead Germans in a Trench' (1918) (Figure 9.2) gives a flavour of the theatricality produced by this technique as the two long-dead German soldiers appear as if lit with artificial light beams, occupying the base of the trench and conveying the stark demarcation of a world divided by life and death.

During this period Orpen produced at least 18 oil paintings of the summer Somme landscape and although '[t]he evidence of death was all around him . . . so was the evidence of life. Skulls and flowers were side by side. . . . Death is even more inscrutable in the face of beauty' (Arnold 1981: 320). The tension between the aesthetics of nature and the ugliness of warfare characterized much of his work in this period.

As the war wore on, 125 of his war works were exhibited at Agnew's gallery in London in May 1918, to considerable popular acclaim but mixed critical response. Some commentators regarded them as lacking sufficient sentiment, drama or action. It was during this exhibition that Orpen offered to donate all his war paintings, and any future ones he completed, to the government under the proviso that they be kept as a single collection. While the exhibition travelled to Manchester and to the United States, any thoughts of showing the paintings in Dublin's National Gallery were shelved amidst fears about how such war paintings might be received in the aftermath of the 1916 Easter Rising. This fear was exacerbated by the artist's close personal friendship with Colonel Lee who oversaw the execution of the rebellion's leaders in Dublin (Arnold 1981; Dark and Konody 1932).

Suffering from continual ill health, possibly from syphilis, and becoming increasingly depressed by the war, Orpen returned to France in July 1918. During this sojourn Orpen's work moved from the realism of his earlier paintings towards a more symbolic or allegorical approach to the conflict. In 'The Mad Woman of Douai' (1918) (Figure 9.3), perhaps one of his most disturbing war paintings, he evokes the destructive impulse on the civilian as well as the military population ushered in by the war.

In the midst of a devastated landscape sits a woman, seemingly having lost her reason, as wearied soldiers and local villagers appear either incapacitated or disinclined to comfort her. The image represents the aftermath of a rape, a metaphor for German brutality that had circulated throughout the war. The violation, however, resides not solely in the body of the woman herself but also on the ruined countryside enveloping the group. The war-weary soldiers occupying the space display no appetite for sympathizing with the woman's experience as they too, perhaps, have endured the ravaging of their bodies and minds over the course of four years of conflict. The psychological as well as the corporeal cost of war, depicted in this painting, represents as increasingly wide recognition that the war's effects were mental as well as physical (Bourke 1996; Leese 2014; Meyer

Figure 9.3 William Orpen, 'The Mad Woman of Douai', 1918.

Source: © Imperial War Museum (Art. IWM ART 4671).

2012; Zuckerman 2004). The affective response to the war was not only felt by the grieving families of dead soldiers but also the emotional toll permeated civil society in wider ways and infiltrated the heads of soldiers who managed to survive the conflict. War art, as well as creative writing, attempted to capture some of these effects, in word and image, and, as Paul Fussell (2000) has argued, fostered a modernist irony in young men 'revealing exactly how spurious were their visions of heroism, and – by extension – history's images of heroism' (Gilbert 1987: 201). The high incidence of shell-shock, estimated as 40 per cent of casualties in the war zones by 1916, challenged earlier interpretations of male hysteria. As Showalter (1987: 63) explains: 'This parade of emotionally incapacitated men was in itself a shocking contrast to the heroic visions and masculinist fantasies that had preceded it in the British Victorian imagination.' Painting, as well as poetry, memoir and novel, subverted these traditional visions and presented uncomfortable but resonating engagements with the effects of modern, mechanized warfare to popular audiences.

After the war's end Orpen remained in Paris, commissioned to document the peace negotiations on canvas, and culminating in the controversial memorializing painting 'To the Unknown British Soldier in France' (1921–1928). Initially he methodically painted the principal politicians and servicemen involved in the conference, a total of 36 figures, gathered in the luxurious surroundings of the Hall of Peace. He then, without notifying the War Museum, erased them all and replaced them with a coffin draped in the Union Jack and guarded by two semi-nude soldiers and cherubs in the air above. This painting was exhibited at the Royal Academy in 1923. Ridiculed by the establishment and the conservative press, the painting was hailed a triumph by the public who voted it the picture of the year. The left-wing press concurred with such a view, with the *Daily Herald* claiming it 'a magnificent allegorical tribute to the men who really won the war' (quoted in Gough 2010: 196). The Trustees of the Imperial War Museum rejected it though for inclusion in its collection of war art, and this prompted the *Liverpool Echo* to opine: 'Orpen declines to paint the floors of hell with the colours of paradise, to pander to the pompous heroics of the red tab brigade' (quoted in Gough 2010: 197). The painting represented an embodiment of the heroism of the soldier in the face of a futile conflict and seemed to touch a nerve among popular English audiences in the immediate post-war period where the memory of the conflict was still fresh. In 1928, however, to mark the death of his friend Earl Haig, Orpen erased the soldiers, the cherubs and the floral tributes and left only the coffin, the gilded marble façade framing it and the beam of light leading to the cross in the painting's background. It was accepted by the Imperial War Museum. After the war Orpen resumed his successful career as a portrait painter with studios in London and Paris, earning a large fortune from his commissions. However, with a chaotic personal life, heavy drinking and continual bouts of ill-health and depression, Orpen's final years were spent often estranged from his family and friends, and he died on 1931 and was buried in Putney Vale cemetery. A hugely successful artist during his lifetime, his work entered relatively obscurity in the aftermath of his death, only to be resurrected by retrospective exhibitions from the 1970s onwards including two staged in the National Gallery of Ireland in 1978 and 2005 (Arnold 1981). William Orpen initially confronted life on the Western Front with some enthusiasm and with the imprimatur of key establishment figures. His talent as a portrait painter at home was mirrored in his early paintings of important military leaders. As the war progressed, however, his approach became darker, as he attempted to capture the landscape at the front as experienced by ordinary soldiers. While his war work did not achieve the critical acclaim of some other official artists, the impact of the conflict on his painting, and on his health more generally, would have enduring effects that proved more potent than any patriotic affiliations to the place of his birth.

Stuck on the home front: John Lavery, 1856–1941

In contrast to William Orpen, John Lavery, the son of a struggling wine and spirit merchant, was born in 1856 in North Queen Street, Belfast and baptised in the St Patrick's Catholic Church, Donegall Street. Three years later his father Henry Lavery, on a voyage on board the American vessel the *Pomona*, was drowned when it struck a sandbank off the coast of Wexford. His wife Mary Donnelly died shortly afterwards from the shock of her husband's death, leaving Lavery and his brother and sister orphans. He was sent to live with his uncle on his farm near Moira, County Down before moving, at age 10, to more prosperous relatives in Saltcoats, Ayrshire. He was unhappy there though and ran away to Glasgow before being returned to his uncle's farm in Moira. When he was 15 years old he departed again for Glasgow with £5 in his pocket, but having experienced some of the rougher sides of Glasgow life he initially moved back with his relatives in Ayrshire.

His creative talents began with a three-year apprenticeship at J.B. McNair's photographic studio in Glasgow where he developed his drawing skills as he touched up negatives and colour prints. In 1874 he started taking classes at the Haldane Academy of Arts in Glasgow with the aim of becoming a portrait artist. Over the next few years he continued to work with photographers and when a studio he was renting was gutted by fire, he received £300 from the insurance company. This prompted him to move to London where he enrolled in 1879 in the Heatherley School of Art before moving to Paris to study at the Académie Julian (Snoddy 2010). It was in Paris that some of his early paintings were first exhibited and where he met James McNeill Whistler, before returning to Glasgow and becoming associated with the emerging Glasgow School of artists, noted in particular for cultivating impressionist and post-impressionist techniques (Billcliffe 2009). In a stroke of good fortune he obtained a prestige commission in 1888 to record Queen Victoria's visit to the International Exhibition in the city. The painting was deemed a success and provided the platform for him becoming a significant society painter. Following the completion of the commission he travelled with friends to Morocco and in 1889 he married Kathleen McDermott, a local flower-seller. They had one child, Eileen, but his wife died of tuberculosis shortly afterwards in 1891.

In the years following, Lavery travelled extensively across Europe and his work was gaining wider recognition among continental artists and galleries. Moreover his reputation as an artist was also gaining traction in Britain and Ireland with his election to the Royal Scottish Academy in 1896 and the Royal Hibernian Society a decade later. With his increased public recognition as an artist he moved to 5 Cromwell Place, Kensington in 1896, and this address would remain significant as his home and studio for some of the most well-known works he produced over the remainder of his life (Snoddy 2010).

Having initially met Hazel Martyn Trudeau – an American artist, socialite and heiress of Irish descent – in 1903, she and Lavery married in 1909 in a union that would see her become the most significant model/muse of his artistic career. She appears in over four hundred of his paintings not only as the subject of portraits but also as the model for other works. 'The Artist's Studio' (1910–1913), produced just before the war, features Hazel, Eileen (John's daughter from his first marriage) and Alice (his stepdaughter), and, like some of Orpen's work, exhibits the continued influence of Velazquez's 'Las Meninas' in early twentieth-century art. The painting also underlines the significance of Hazel to his career and, despite a tempestuous marriage, she would continue to exert a long-term influence on his work (McCoole 1996). Outside the Kensington studio, he undertook trips to Tangier and other North African locations between 1910 and 1914 and painted the North African landscape and its people. Kenneth McConkey (2010: 117) claims that through such trips '[c]olourfully clad Moors, Derbers and Nubians became the painter's antidote to society ladies, who habitually arrived in the studio [in London] dressed in black and grey'. Tours to Venice and Wengen where he also painted *en plein air* provided an additional escape from the routine of portraiture that occupied much of his time at his London studio.

In 1913 the Laverys visited Killarney House, in County Kerry, beside the lakes, mainly to paint a portrait of Lady Dorothy Browne. But it was here that John Lavery initially started to develop his idea for his triptych 'The Madonna of the Lakes' (1917) featuring Hazel as the Madonna, Alice as a young St Patrick and Eileen as St Brigid. A backdrop of the Killarney mountain landscape and its lakes connected the three sections of the composition. Edwin Lutyens, a friend of John Lavery, was commissioned to design the frame using Celtic spiral motifs. When exhibited at the Royal Academy in 1919 it was hailed as 'an enormous advance on anything to be seen in modern religious painting' (quoted in McConkey 2010: 130). Lavery donated it to the church in Belfast where he was baptised and it consolidated his wife's iconic status in his work. The painting's execution during the years between the start of the Great War and the Easter Rising illustrates the complex loyalties the family had between his success as a well-respected portrait painter in London and their strong affiliations to Ireland and its nationalist movement. Indeed the Laverys spent increasing time in Ireland and at the outbreak of the war in 1914 they were visiting Dublin and Wicklow. Hazel Lavery's Galway ancestry, alongside her husband's Catholic Belfast roots, prompted him 'to ponder more deeply the question of national allegiance' (McConkey 2010: 124). Moreover their social circle in Dublin brought him into direct contact with many of the major figures spearheading the agitation for Home Rule.

However on their return to London in the summer of 1914 the impact of the war was immediately apparent to them. Lavery set about depicting the war's destructive force in his early painting 'The First Wounded, London Hospital 1914' (Figure 9.4) (Park and Park 2011).

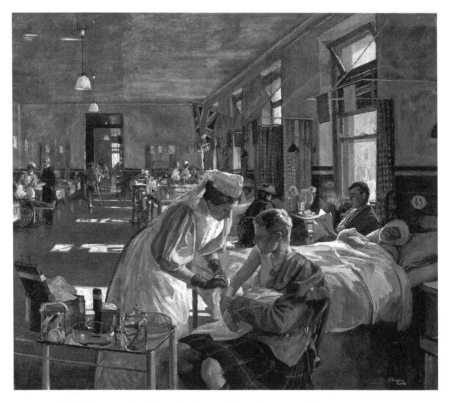

Figure 9.4 John Lavery, 'The First Wounded, London Hospital 1914'.

Source: Photograph courtesy of the Dundee City Council, Dundee's Art Galleries and Museums.

The picture was exhibited at the Royal Academy, and won critical and popular acclaim from a public increasingly hungry to see the effects of the war at home. The composition, set in a London military hospital, appealed to a large audience because it conveyed the immediate wounds of war and domesticated the conflict for those at some distance from the sites of battle. Moreover it also provided an insight into the medical and nursing care provided to these early casualties of the war. Unlike Orpen, Lavery's chronicling of the Great War was to be from the perspective of the home front (Cooksley 2006; Grayzel 2002). He joined the Artists' Rifles in 1915 but because of his age and health was advised by his doctor that his contribution to the army should reside in visually depicting the conflict rather than fighting in it. Initially he continued with his portrait commissions, particularly of political leaders including Winston Churchill, John Redmond [leader of the Irish Parliamentary Party] and Edward Carson [leader of the Ulster Unionists]. His immediate attitude to the Easter Rising in April 1916 is difficult to discern, but it is known that he supported the idea of introducing conscription to Ireland. At the behest of the Lord Chief Justice Sir Charles Darling, he also recorded the Court of Appeal's hearing for Roger Casement's conviction for high treason and sentencing to death in 'The Court of Appeal' (1916–1918).

While Lavery was anxious to travel independently to the Front to witness the war first-hand, under military restrictions there was no possibility of him securing permission. While he spent some time in early 1917 in St Jean de Luz and painted its harbour, he was never to get close to the action at the front lines and consequently his war paintings were to evoke conditions on the home front. In July 1917 21 German Gotha biplanes carried out an aerial bombardment of London, which was clearly visible from Lavery's Cromwell Place studio. He captured the scene on canvas depicting Hazel, with her back to the viewer, kneeling before a statue of the Madonna as the sky outside erupted in the aerial battle between German bombers and British defence forces ('Daylight Raid from my Studio Window', 1917). Depicting this attack on the capital by combining the domesticity of their interior world of home and studio with the aerial war being conducted outside their window conveyed to the public that the conflict was not just conducted in spaces afar across the English Channel but that it was also intimately bound up with the lives of those who remained at home (Chapman 2014).

Lavery was increasingly keen to play an official role in representing the war and with the establishment of the new Department of Information under the stewardship of John Buchan, his opportunity arose to become an official war artist. His desire followed the wider motivation of artists 'to witness, interpret and leave some form of personal testimony was a powerful incentive to those who needed to come to terms with their violent muse' (Gough 2010: 32). However as Lavery now was in his early sixties he was restricted to home-front duty and was issued with a Special Joint Naval and Military Permit. These permits were in reality letters of introduction and did not necessarily give artists freedom to depict any aspect of life on the home front that they liked. He initially went to Scotland to draw the fleet at anchor in the Firth of Forth and, despite some negotiations with local commanders, he made several paintings including 'The Forth Bridge 1917'. The image depicts the bridge itself, kite balloons and the grand fleet at some distance in the background. While much of the work of the official war artists captured the life of the infantry soldier at the front, this image highlighted the significant role of the navy in the prosecution of this conflict. On his way to Edinburgh he stopped at Newcastle-upon-Tyne and visited the munitions factories at Elswick. Lavery decided to portray this interior space providing a pictorial representation of the industrial scale of production underpinning the industrial speed of killing along the battlefronts. In 'Munitions, Newcastle, 1917' (Figure 9.5), the size of the machinery and weaponry of war dwarfs the workforce responsible for their production.

The high ceilings, the vast girders and the armaments themselves convey a sense of the technological and human effort invested to provide the infrastructure of what would later be described as the first modern war. The painting also highlights the transformation in gender roles that the war precipitated as women became centrally involved in the manufacture of weapons (Ouditt 1994; Woollacott 1994).

Figure 9.5 John Lavery, 'Munitions, Newcastle, 1917'.

Source: © Imperial War Museum (Art.IWM ART 1271).

In Southampton, later in 1917, Lavery provided a bird's eye view of soldiers embarking on ships heading for the front lines – 'Troops Embarking at Southampton for the Western Front, 1917'. The size of the ship and the cranes surrounding it renders diminutive the long lines of smartly uniformed soldiers preparing to board the vessel. The hues of brown, khaki and grey convey something of the colour of the landscape that would emerge in the paintings of artists working

Figure 9.6 John Lavery, 'A Convoy, North Sea, 1918. From N.S. 7'.

Source: © Imperial War Museum (Art.IWM ART 1257).

at the front. Lavery was dispatched north again, including a visit to the Orkneys, where he could experience first-hand the severity of the winter weather conditions under which the Royal Navy was operating. In a visit to East Fortune, in East Lothian, Lavery conducted the preparatory work for 'A Convoy, North Sea, 1918. From N.S. 7' (Figure 9.6).

Rather than imagining the scene from land, Lavery, now 62 years old, sketched the scene from an airship and evoked the danger faced by gunners leaning out over the sea in search of U-boats. Moreover the scene indicates the geopolitical significance of keeping the North Sea open for military and commercial convoys. His final commission, as the war ended in 1918, was to provide the Imperial War Museum with a series of paintings for their Women's Work collection, including the 'The Cemetery, Etapes' (1919). Lavery's contribution to the corpus of war art earned him a knighthood.

In the years after the war Lavery's connections to Ireland deepened as he produced numerous paintings of politicians and churchmen, on both sides of the political divide. The family established a strong friendship with the Nationalist leader Michael Collins, and they provided their London house as a retreat for the Irish delegation that arrived in England to negotiate the terms of the Anglo-Irish Treaty. After independence Lavery continued his close connections with Ireland. He completed several portraits of Irish Free State politicians, and donated many of his works to the Hugh Lane Municipal Gallery in Dublin and the Belfast Museum. Moreover Hazel Lavery served as model for her husband's commission to produce an allegorical image of Ireland, which was subsequently used on Irish banknotes between 1928 and 1975, and became the watermark on Euro notes introduced in 2002. Lavery died in 1941 in Kilkenny where he was guest of his stepdaughter and was buried beside his wife in Putney Vale cemetery.

Afterlife: in the shadow of centenaries

The contribution of two Irish painters to immortalizing the war effort on canvas is beginning to attract more attention in discussions of Ireland's role in the Great War. The two men shared certain similarities, most notably, enjoying successful careers in England, as portrait artists, with strong connections with the social elite of the day. Both were also keen to play a role in the war and the official war artists' scheme afforded them such an opportunity. Their difference in age, however, coupled with Orpen's significant contacts with the political establishment, meant that he would paint at the war front while Lavery would be confined to depicting the war's impact at home. They both produced portfolios of work that contained many arresting images of the human and physical costs of war, and each, in their own way, spoke to the emotional and moral questions raised by four years of conflict. They were both rewarded with knighthoods. That said, the Dubliner Orpen, who had many personal connections with Irish nationalists, over time, distanced himself from the political turmoil that enveloped Ireland during the war years. His retreat in the post-war years from family and friends, both in England and Ireland, perhaps is emblematic of the significant long-term impact serving as a war artist on the front had on his psychological and physical well-being. By contrast, the Belfast man Lavery, maintained a longer and deeper relationship with the island of his birth, and while his ultimate political views remain below the surface, the legacy of his artistic output indicates a highly nuanced set of geographies of allegiance. It is precisely the complexity of these loyalties, mirrored in his paintings, that renders his work, emblematic of the entangled topologies of identity and memory, prevalent on these islands from the First World War to the present. Both of these artists performed the war through visual rather than verbal media. The communication of life in the trenches and at home, under wartime conditions, was translated through the vocabulary of the painterly image, evoking landscapes of death and domesticity that reflected both their personal experience and the wider geographies of identity each held. The emotional registers that the Great War provoked for each painter speaks to the deeper complexities of their ideological and affective commitments to Ireland, before, during and after the First World War.

References

Anderson, B. 2013. Affect and emotion. In N.C. Johnson, R.H. Schein and J. Winders (eds) *The Wiley-Blackwell Companion to Cultural Geography*. Oxford: Wiley-Blackwell, 452–464.

Arnold, B. 1981. *Orpen: Mirror of an Age*. London: Jonathan Cape.

Billcliffe, R. 2009. *The Glasgow Boys*. London: Frances Lincoln.

Bourke, J. 1996. *Dismembering the Male: Men's Bodies, Britain and the Great War*. Chicago: University of Chicago Press.

Chapman, T. 2014. *IWM: The First World War on the Home Front*. London: Andre Deutsch.

Cooke, S. and Jenkins, L. 2001. Discourses of regeneration in early twentieth-century Britain: from Bedlam to the Imperial War Museum. *Area* 33: 382–390.

Cooksley, P. 2006. *The Home Front: Civil Life in World War One*. London: The History Press.

Dark, S. and Konody, P.G. 1932. *Sir William Orpen: Artist and Man*. London: Seeley Service.

Eksteins, M. 2000. *Rites of Spring: the Great War and the Birth of the Modern Age*. New York: Mariner Books.

Evans, R.J.W. 2014. The greatest catastrophe the world has seen. *New York Review of Books*, 6 February.

Foster, R. 2005a. 'Old Ireland and Himself': William Orpen and the conflicts of Irish identity. *Estudios Irelandeses* 1: 39–50.

Foster, R. 2005b. Orpen and the new Ireland. In R. Upstone (ed.) *William Orpen: Politics, Sex and Death*. London: Philip Wilson, 63–155.

Fussell, P. 2000. *The Great War and Modern Memory*. Oxford: Oxford University Press.

Gilbert, S. 1987. Soldier's heart: literary men, literary women, and the Great War. In M. Higonnet, J. Jenson, S. Michel and M. Collins Weitz (eds) (1989) *Behind the Lines: Gender and Two World Wars*. New Haven: Yale University Press, 197–226.

Gordon, D. and Osborne, B. 2004. Constructing national identity in Canada's capital, 1900-2000: Confederation Square and the National War Memorial. *Journal of Historical Geography* 30: 618–642.

Gough, P. 2010. *A Terrible Beauty: British Artists in the First World War*. Bristol: Sansom and Co.

Grayson, R. 2010. The place of the First World War in contemporary Irish republicanism in Northern Ireland. *Irish Political Studies* 3: 325–345.

Grayzel, S. 2002. *Women and the First World War*. London: Routledge.

Harries, M. and Harries, S. 1983. *The War Artists: British Official War Art of the Twentieth Century*. London: Imperial War Museum.

Heaney, S. 2004. *This Week*, 15 April.

Higonnet, M., Jenson, J., Michel, S. and Collins Weitz, M. 1989 (eds) *Behind the Lines: Gender and Two World Wars*. New Haven: Yale University Press.

History Ireland. 2014. Ireland and WWI. Special issue of *History Ireland* 22: 3–66.

Hynes, S. 1990. *A War Imagined: the First World War and English Culture*. London: Bodley Head.

Jeffrey, K. 2000. *Ireland and the Great War*. Cambridge: Cambridge University Press.

Jeffrey, K. 2014. William Conor's People's War: a consideration of the most 'Irish' of Ireland's First World War painters. *History Ireland* 22: 42–44.

Johnson, N.C. 1999. The spectacle of memory: Ireland's remembrance of the Great War 1919. *Journal of Historical Geography* 25: 36–56.

Johnson, N.C. 2002. Mapping monuments: the shaping of public space and cultural identities. *Journal of Visual Communication* 1 (3): 293–298.

Johnson, N.C. 2007. *Ireland, the Great War and the Geography of Remembrance.* Cambridge: Cambridge University Press.

Kavanagh, G. 1988. Museum as memorial: the origins of the Imperial War Museum. *Journal of Contemporary History* 23: 77–97.

Keating, S. 1937. William Orpen: a tribute. *Ireland Today*, n.p.

Kidd, W. and Murdoch, B. (eds) 2004. *Memory and Memorials: the Commemorative Century.* Aldershot: Ashgate.

Leese, P. 2014. *Shell Shock: Traumatic Neurosis and the British Soldiers of the First World War.* Basingstoke: Palgrave Macmillan.

McConkey, K. 2010. *Sir John Lavery: a Painter and his World.* Edinburgh: Atelier (second revised edition).

McCoole, S. 1996. *Hazel, a Life of Lady Lavery.* Dublin: Lilliput Press.

Malvern, S. 2000. War, memory and museums: art and artifact in the Imperial War Museum. *History Workshop Journal* 49: 177–203

Malvern, S. 2004. *Modern Art, Britain and the Great War.* New Haven and London: Yale University Press.

Meyer, J. 2012. *Men of War: Masculinity and the First World War in Britain.* Basingstoke: Palgrave Macmillan.

Orpen, W. 1921. *An Onlooker in France 1917–1919.* Fairford: Echo Library 2010 edition. Originally published London: Williams and Norgate, 1921.

Ouditt, S. 1994. *Fighting Forces, Writing Women: Identity and Ideology in the First World War.* London: Routledge.

Park, M.P. and Park, R.H.R. 2011. Art in wartime: 'The First Wounded, London Hospital, August 1914'. *Medical Humanities* 37: 23–26.

Pennell, C. 2012. *A Kingdom United: Popular Responses to the Outbreak of the First World War in Britain and Ireland.* Oxford: Oxford University Press.

Sanders, M.L. 1975. Wellington House and British propaganda during the First World War. *The Historical Journal* 18: 119–146.

Showalter, E. 1987. Rivers and Sassoon: the inscription of male gender anxieties. In M. Higonnet, J. Jenson, S. Michel and M. Collins Weitz (eds) (1989) *Behind the Lines: Gender and Two World Wars.* New Haven: Yale University Press, 61–69.

Snoddy. T. 2010. 'Sir John Lavery'. *Dictionary of Irish Biography.* Accessed online: http://dib.cambridge.org/viewReadPage.do?articleId=a4701, last accessed 8 October 2015.

Taylor, P.M. 1999. *British Propaganda in the 20th Century.* Edinburgh: Edinburgh University Press.

The Times. 1916. 20 May.

Turpin, J. 1979. William Orpen as student and teacher. *Studies: an Irish Quarterly Review* 68: 173–192.

Upstone, R. (ed.) 2005. *William Orpen: Politics, Sex and Death.* London: Philip Wilson.

Weight, A. 1986. The Kensingtons at Laventie: a twentieth century icon. *Imperial War Museum Review*: 14–18.

Winter, J. 1995. *Sites of Memory, Sites of Mourning.* Cambridge: Cambridge University Press.

Winter, J. 2003. *The Great War and the British People.* Basingstoke: Palgrave Macmillan.

Woollacott, A. 1994. *On Her Their Lives Depend: Munitions Workers in the Great War.* Los Angeles: University of California Press.

Zuckerman, L. 2004. *The Rape of Belgium: the Untold Story of World War I.* New York: New York University Press.

Part III

Commemorative vigilance and rituals of remembering in place

10 Embodied memory at the Australian War Memorial

Jason Dittmer and Emma Waterton

Much of the literature that deals with what Jay Winter (2006) labels 'historical remembrance' makes clear reference to the issue of identity – particularly *national* identity. As Martin Heisler (2008a, 2008b; see also White 2000) explains, this is because processes of national-identity formation often look for a moment in the past that can be fleshed out with testimonies of survival, sacrifice or overcoming significant hardships. Borrowing James Wertsch's (2008a: 60; see also 2008b) terminology, these can be described as 'schematic narratives' that simplify complex histories and efficiently organize – through repetition – how a society (collectively) ought to 'think about' and respond to a nation's past. As a number of scholars highlight, museums and memorials, along with a host of other heritage sites, are often utilized in the construction and maintenance of these sorts of memories (Smith and Waterton 2009; Sather-Wagstaff 2011; Macdonald 2013). Less often, the focus of debate is on the ways in which these collective memories articulate with us as individuals and settle into our own personal worlds as feelings and affects. To remedy this, our chapter picks up a task originally cast in the work of Geoffrey White (2000: 505), whose aim was to examine the 'practices that traverse the "out there" of collective representation and "in here" of personal cognition and emotion' for visitors to the US national memorial to the Pearl Harbor attacks.

In parallel to White, this chapter takes the Australian War Memorial (henceforth 'the Memorial') as a focal point, foregrounding a particular instance of historical remembrance: that concerned with the Kokoda campaign of the Second World War. Alongside Gallipoli, the Kokoda campaign is frequently called upon as foundational fodder for Australia's collective memories, serving as a reminder of the quintessential national character. While much has been written about the relationship between Kokoda and the Australian imaginary – along with the work it does in representing contemporary Australian identity – less scholarly attention has been channelled towards understanding it beyond its representational nature. In this chapter, then, we start from the premise that Kokoda is not a representation in and of itself, confined to the past; rather, to borrow from Curti (2008: 108), it is 'a continually embodied process working towards a future'.

With this in mind, we have parcelled our analysis into two parts. First, we look to the interpretive strategies employed at the Memorial and identify the schematic narrative template of Anzac (Australian and New Zealand Armed Corps) underpinning the Kokoda campaign's representation. Second, we examine the strategies employed to impassion the narrative, thinking particularly about the range of kinaesthetic senses and flows that may be enacted – as sensations, feelings and atmospheres. The evocative power of such strategies lies in the invitation they extend to visitors to *feel* something as they reflect on the horrors of war. These registers of affect are woven into the fabric of the Kokoda exhibit; while they may not always be expressible, our contention is that the exhibit has been designed to trigger deeply felt, physical and visceral responses that emerge as visitors react to atmosphere and, oftentimes given the particular context of the Memorial, a sense of haunting (O'Riley 2007; Coddington 2011). Our method entails first considering the larger narrative of Australian militarism, Anzac memory and nationhood, which is rendered visible through commemorations of Gallipoli at the Memorial. We then look to the specific iteration of the narrative that is Kokoda, shifting from the textual register to a more somatic one. First, though, we provide a review of the literatures we have used to construct a sense of an embodied collective memory.

Embodying collective memory

'Collective memory' has achieved a great deal of academic traction in the last quarter century (Middleton and Edwards 1990; Connerton 1991; Halbwachs 1992; Wertsch 2002; Wertsch and Roediger 2008; Olick et al. 2011; Olick 2013). A significant portion of that work has been dedicated to understanding the role played by remembrances of war – in spatial and symbolic terms – within wider processes of national meaning- and identity-making (see, for example, Nora 1989; Muzaini and Yeoh 2005). The continuing production of scholarly work dedicated to these themes suggests that it has clear residual value and thus will remain a perennial concern. But this ought not to imply that it is immune to conceptual massage: far from it, the study of collective memory remains a cumulative research enterprise, to which we hope to add an embodied and yet more-than-human interpretation that takes the term itself as a rather literal starting point. Indeed, 'collective memory' has been the subject of some controversy given the way in which it modifies the term 'memory' with the adjective 'collective' (Bartlett 1995), implying that collective memory functions in the same way as an individual's memory. Wertsch (2008b: 121) notes that this 'strong' version of collective memory, in which the group literally holds memories, has usually been rejected (see also Wertsch and Roediger 2008). He argues instead for an instrumental and distributed notion of collective memory, which 'involves agents, acting individually or collectively, and the cultural tools they employ, tools such as calendars, the Internet, or narratives' (Wertsch 2008b: 121). Thus, collective memory is something that is worked out between active agents who remember together (Halbwachs 1992), and as such it is a political process that is constantly being re-negotiated.

As Wertsch (2008b: 121) goes on to argue,

[t]his does not mean that such memory somehow resides in texts or records, but it does mean that with the rise of new forms of external symbolic storage such as written texts or the Internet, the possibilities for remembering undergo fundamental change.

This upends the colloquial understanding of remembrance as an internal process *within* the subject; memory is itself dependent on the materials that are deployed as tools to enhance recall. So, to put it one way, *all* memories are collective in that they emerge from the interplay of bodies and the various memory-tools around them, which are themselves shared and distributed – often by semiotic means (Halbwachs 1992; Wertsch 2002; Wertsch and Roediger 2008; Waterton and Watson 2014). Different memory-tools will augment different forms of memory and (re)shape practices of remembering around particular events, objects or people (Wertsch and Roediger 2008). The material properties of such memory-tools become important – not in and of themselves, but in their interactions with other objects and sensing bodies (Massumi 2002). This approach highlights both the political nature of memory-work and also the political agency of the memory-tools themselves, which can bring forth unbidden a range of histories. Here, we can see points of intersection with literature on the geography of heritage and memorialization, which has always emphasized the role of material objects and landscapes in the production of heritage. For instance:

> Memorials influence how people remember and interpret the past, in part, because of the common impression that they are impartial recorders of history. Their location in public space, their weighty presence, and the enormous amounts of financial and political capital such installations require imbue them with an air of authority and permanence. . . . Further, their apparent permanence suggests the possibility of anchoring a fleeting moment in time to an immovable place. Composed of seemingly elemental substances – water, stone, and metal – memorials cultivate the appearance that the true past is and will remain within reach.
>
> (Dwyer and Alderman 2008: 167–168)

The desire to 'reach' back in time and make political use of the past in the present is a common understanding of memory, perhaps best immortalized in George Santayana's truism 'those who cannot remember the past are bound to repeat it', or, with a stronger political inflection, George Orwell's often-quoted observation, '[w]ho controls the past . . . controls the future: who controls the present controls the past' (1949: 88). The desire to shape practices of memorialization, then, is arguably one that has been with us for some time and is often underpinned by intent to shape future political action. The production of collective memory is no different and can be understood as a political attempt to conjure up the past and collectively remember an event in a particular way.

While this chapter starts from a similar premise about the political nature of memorialization, we advance that line of thinking by offering two related (friendly) critiques, drawing on the insights of Henri Bergson (1988), Manuel DeLanda (2006) and Gilles Deleuze (1991). Together these thinkers share a relational ontology that superficially resembles the instrumental and distributed form of collective memory that is outlined by Wertsch. However, their work pushes us to consider collective memory in slightly different ways. The first critique is of the 'subject' within this formulation. DeLanda (2006) and Deleuze (1968, 1991) both emphasize the emergence of human subjectivities out of the flux and flow of material conditions. If we consider the subject as dependent on material context for its particular form (e.g. a particular body, sustenance, atmospheres and so forth) then it becomes clear that the subject is destabilized as an active, unproblematic agent. Rather, the subject and its agency are constantly becoming, and becoming differently, as processes of assemblage unfold (Protevi 2009; see also Grosz 2005). So who, then, is doing the remembering? Or, put differently, how does the act of remembering re-work the subject (individual or collective)? We know that, at the scale of the individual, remembering is crucial to the continued production of the self and its agency. The absence of memory is something that – when taken away by cruel medical conditions such as Alzheimer's disease – can be profoundly destabilizing.

We also know that, at the scale of the collective, remembering is crucial to the continued production of the 'we' and 'our' agency, as well as narratives of togetherness (Waterton et al. 2010). When taken away – perhaps through colonial forms of education – limits are put in place that reduce the ground from which anti-colonial practices can emerge (see contributions to Dei and Kempf 2006). Of course, under most circumstances the contestation of memory is not so existential; that is, conflict emerges not over whether there is a narrative but what that narrative is. This is a point we will return to in our discussion of narratives, but suffice to say here that the recognition that memory is not just in our minds but is distributed through various media is but a step towards a further truth: the political subject is distributed well beyond our bodies as well. The active agents in the distributed model of collective memory are not just co-producing memories, they are co-producing themselves, in conjunction with the memorials that they have both inherited and produced.

Our second critique draws from Bergson's (1988; see also Deleuze 1991) work on time and becoming. Drawing on the above discussion of the distributed subject, if political action is predicated on the assemblage that enables subject formation in any given moment, then how can we account for memory? A common assumption of collective memories is that they are reservoirs of insight and experience that we can go back to in order to make decisions. Bergson, however, noted that time's arrow only went one way, in contradiction to much of the philosophy of time that was emerging contemporaneously from the physics of relativity (1988). This led to Bergson's position on memory, which was that remembering was not a process of going back in time, but rather of bringing the past into the present. This may seem like pedantry, but only the latter was congruent with time's arrow. In Deleuze's (1991) adaptation of Bergson's ideas, he conceptualized memory as but one of many virtual entities, such as imagined events, that could be made co-present in the now and that might shape political cognition and action. In other words, brought forth from the past, memories shape our actions in the present (Bergson 1988; Deleuze 1991). Given the materialist ontology of assemblage, this means that memory must be materialized in some way that enables it to enter into assemblage. In some ways, this is easy to address: experiences can lay down a somatic marker within the body (Connolly 2002) – a habit, perhaps – which is activated by related sensory experiences. Think of, for instance, the way in which your body jerks away from grabbing a hot skillet (again) without conscious thought. The memory of past burns is materialized within the neurological system, waiting to be activated by a parallel embodied experience. Bergson recognized this as the past being brought into the present when he wrote of habitual remembering, which he argues 'no longer *represents* our past to us, it acts it; and if it still deserves the name of memory, it is not because it conserves bygone image, but because it preserves their useful effect into the present moment' (Bergson 1988: 81–82). Of course, 'real' events are not the only source of somatic markers. A near-miss can have the same adrenaline-soaked embedding within the self: for example, the time a momentarily forgotten child wanders too close to the cliff's edge. In a flash, an imagined future is conjured forth that can shape future behaviour in an attempt to stave off the actualization of that future. In a more positive iteration, the promise of a utopian future can equally saturate political rationality. What links all these virtualities is that they involve the locating of memories within the human body (but in conjunction with other matter), where they are sensed and felt in ways that are productive of particular subjectivities (Crang and Tolia-Kelly 2010; Waterton and Watson 2014). As Amanda Kearney (2009: 213) articulates it: '[T]he depository of these intangible cultural expressions [i.e. collective memory] is the human mind and body, ancestors, and homelands; all of which become instruments for its enactment, or literally its embodiment.'

It is worth noting that Bergson's materialist critique of instrumental and distributed collective memory can go one step further. While 'things' can enter into memory assemblages as media, inscribed by human activities to produce memorials, they are also subject to the power of time, independent of any human practices of memorialization. That is, they are subject to processes of erosion and decay or, alternatively, self-organization. While some of these 'texts' only take on meaning when considered through the discourses of scientific knowledge (for instance, the 'history' of a rock or a landscape), others have an extra-discursive dimension that is not reducible to human structures of language and meaning. DeLanda (2006) highlights the example of DNA as an encoding of living assemblages that exceeds our efforts to tell stories about it (although we can certainly tell those stories, as numerous genealogy/DNA analysis websites offer).

Taken together, our two critiques offer a model of collective memory that sees individual and collective subjects becoming together, constantly de- and re-territorializing around new material bodies, objects and contexts, which alter not only *what is remembered*, but *who is doing the remembering*. These two questions are fundamental for demonstrating how high the political stakes around memory can become. In the following section, we look to the technologies that are brought to bear in efforts to shore up the stability of a particular collective subject. Indeed, it is precisely because collective memory is reliant on processes of materialization that it is subject to governmental intervention.

Materializing narrative

Collective memory is a powerful force for shaping bodies politic at multiple scales (on bodies politic, see Protevi 2009). The relationship between collective memory and narrative is not straightforward. Harnessing our earlier point that memory need not be discursive in nature, but may operate beyond the realm of discourse (recall the example of DNA), it is productive to think of collective memory as rich in potential, drawing from all possible constellations of bodies, materials and objects (actual or virtual), and capable of influencing many human and more-than-human processes. Narrative is the result of the actualization of these resources within discourse. However, narrative is not simply the result of collective memory; rather, it acts back on the processes of collective memory. Wertsch quotes Bruner (Wertsch 2002: 89) to the effect that narrative is 'our preferred, perhaps even our obligatory medium for expressing human aspirations and their vicissitudes, our own and those of others. Our stories also impose a structure, a compelling reality on what we experience, even a philosophical stance'. Therefore, the forms of narratives are themselves virtual elements of any collective memory assemblage.

Without adopting a materialist ontology, Wertsch (2008b: 122) offers a relational approach to narrative that parallels our own interest in the ontology of assemblage:

> What I have in mind derives from the fact that the textual resources used to produce narratives invariably have a history of use by others. Paraphrasing Mikhail Bakhtin (1981, 293), this means that narratives are always half someone else's, and it leads to questions about how narrators can coordinate their voice with those of others that are built into the textual resources they employ.

Narratives, then, are poly-vocal, bearing the marks of the past. This opens up some opportunities even as it closes others down. Further, the narrative is not just *told* by a speaker, but that speaker must work to (re)construct the narrative as well. Wertsch focuses on 'schematic narrative templates', which he differentiates from specific narratives in that they have a meta-quality. In the field of critical geopolitics, these meta-narratives have been termed 'plots' (Dittmer 2010), in which particular roles in the narrative can be filled by a range of actual protagonists, or in a more abstract sense it is possible to think of genre as a kind of schematic narrative template. These may be considered in Deleuzean (1968) terms as a form of 'difference and repetition', in which individual instantiations of the cultural form resemble each other because of similar conditions of production, yet differ enough from one another to enable narrative evolution over time (for an example of this narrative tension between constancy and change, see Dittmer 2013).

In Wertsch's empirical example of schematic narrative templates – Russian narratives of the Second World War – he sees a range of authorial voices contributing to the individual narratives: school curricula (rapidly changing in the post-Soviet context); the speaker's friends; and the individual speaker. The empirical arc that he presents is of declining specificity in the narrative of the Second World War offered; still, he is at pains to note that even the most bare-bones narrative offered (that of a 15-year-old educated entirely in the post-Soviet era) maintains the schematic narrative template offered by the longer essay: Russia was innocent, and repulsed an invader. This template, Wertsch argues, underpins much of Russian history and serves to organize the various elements of Russian history into an iterative narrative. This iterative narrative itself serves as the basis for the persistence of Russian identities (always plural and yet always cohering) over time and also aids in the normative disciplining of bodies around those identities.

In Wertsch's useful formulation, the schematic narrative template serves as a technology for producing particular national (or other collective) narratives for several reasons. First, the template itself is subject to state power through its influence on school curricula, museum funding and media regulation (there are, of course, limits to this, particularly in less authoritarian states than Putin's Russia, but the point holds to varying degrees regardless of the state). This means that the state can manipulate this template to its own ends, even if this ability is only partial. Second, the template provides a sensibility that filters the flow of potential information into that which can be easily assimilated to the template. In other words, it provides a means to evaluate new information for validity. Walter Fisher (1987) refers to this processing of new information as 'narrative rationality'; rather than comparing new information to empirical benchmarks (*logos*), most information is sorted by how well it fits into already existing narratives with which we identify (*mythos*). Third, as indicated by Wertsch's exemplar (the 15-year-old student), the template is resilient in the absence of information. In fact, if anything, the template is more troubled by increases in the amount of information, through which the level of detail might eventually swamp the ability of the template to accommodate it. It is for this reason that academics argue that history is 'more complicated' than it is often taught, and also that governments tend to not trust academic historians with the national curriculum (see Sellgren 2013; Hartmann 2015).

For these reasons, we find the concept of schematic narrative templates to be compelling. But, as would be expected given the discussion of collective memory above, we think that the concept as it stands is relatively immaterial, and attention has to be directed to the ways in which these templates are somaticized and grounded within particular cultural contexts. For instance, despite the seeming ubiquity of some templates (both nationally and transnationally), there are always other templates: dissident histories, alternative communities, other ways of being and becoming. So what accounts for the visceral rootedness of some templates and the relative ephemerality of others? In order to account for the power of some templates over others, it becomes necessary to look beyond the text to the material forms in which these templates become actualized and circulate. While this may seem contrary to Wertsch's insights, which were gained through the reduction of narratives to meta-narratives, it is a necessary step in order to gain insights into the workings of his concept. After all, one does not encounter narrative templates; one encounters narratives, and it is these narratives that are seen, heard and felt. Indeed, it is this sensory experience of narratives that makes them 'stick' in bodies. Our research at the Memorial can thus be understood as an attempt to consider the ways in which schematic narrative templates are materialized in bodies as a result of those bodies' participation in multiple (and often competing) assemblages of national heritage. Here, the Memorial becomes a site through which a schematic narrative template of Australian heritage is materialized in a range of media, each of which engages with museum-goers' senses in different ways. It is the particularity of these media, and more specifically the particularity of each encounter, that shapes the ways in which heritage becomes embodied.

Embodied heritage: materializing schematic narrative templates

The Australian War Memorial is located in Canberra, Australia, buttressed by everyday markers of remembrance: in nearby street names (Anzac Parade) and adjoining parklands (Anzac Park and the Remembrance Nature Park). A visitor approaching the Memorial today would find the culmination of a 16-year-long vision spearheaded by the journalist and historian Charles E. Bean, who summed up the Memorial in 1948 as such: 'Here is their spirit, in the heart of the land they loved; and here we guard the record which they themselves made.'[1] Completed in 1942, the sandstone complex consists of a cloistered Roll of Honour, the domed Hall of Memory incorporating the Tomb of the Unknown Australian Soldier, the Pool of Reflection, the Flame of Remembrance, a sculpture garden, several galleries including Anzac Hall and the Hall of Valour, a Discovery Zone and other educational spaces, a research centre and a parade ground. Wandering through the complex, visitors can expect to encounter an impressive array of archival resources, such as large relics, artefacts, dioramas, film, photographs, official and private records, diaries, paintings, oral histories and sounds (such as The Last Post ceremony), as well as the names of the fallen.

As an institutionalized – and national – heritage site, the Memorial and its components are already rendered with a particular meaning and significance, articulated through the site's exhibitions, performances and displays. As visitors, we are asked to engage with and relate to those meanings and significances but they can never capture the full story; indeed, they say only very little about personal memories, relationships, energies and gestures. In other words, '[w]hat a particular site "is", and how it *feels*, can become highly variable' (Crouch 2015: 186, emphasis in original). Sometimes, there is a power to *move* and we are drawn into a display, connected. At other times, our experiences carry a more quiescent quality, muted by boredom and fatigue – and we become disengaged, disinterested. Either way, our experiences fall under the rubric of affect, which as Hayden Lorimer (2008: 2) reminds us is distributed in and between bodies, objects, technologies, sounds, activities, relations and any number of other things.

This is a point not lost on those charged with planning and designing sites of memory, who have bolstered traditional textual panels and audio-visual tours with increasingly varied and sophisticated interpretation strategies, many of which are aimed at triggering an affective pedagogy (see Witcomb, Chapter 12, this volume). Thus, while lighting, logistics, sound and visitor 'flow' have for some time been considered powerful elements in creating museum atmospheres, highly mediated technologies are also being called upon to 'trigger and diffuse' affect (Thrift 2008: 254). As we become more and more attuned to the ways in which affective atmospheres and their potentialities are being engineered into places, we simultaneously have to raise our awareness of their transmission and work upon ways of capturing them. This means paying closer attention to how museums themselves – their buildings, settings and internal fixtures, as well as the interpretative devices they contain – provoke a range of experiential and affective potentialities that afford, in turn, all kinds of movements and feelings (after Adey 2008: 441; see also Crang and Tolia-Kelly 2010). Potentiality is a key qualifier here as it implies contingency and context, as David Bissell (2010: 83) makes clear that 'different sets of things, their configuration, their assemblage and spacing, their energy, have different capacities to do different things' with the human capacity for self-reflexivity and intention a key dimension.

Foregrounding the affective capacities of the Memorial immediately brings with it the challenge of figuring out how to access what many scholars have termed 'the unspeakable': swiftly occurring relational responses, intensities and sensuous experiences. The literature that deals with the more-than-representational warns of this difficulty, which Derek McCormack (2002: 470) terms 'the cognitive threshold of representational awareness'. Being mindful of this threshold, our work at the Memorial evolved around the concepts of 'performance' and 'ethnography', a poetics of inquiry, which, to borrow again from McCormack (2008: 2), entailed 'becoming affected and inflected by encounters' within the Memorial's exhibition spaces. We thus used our bodies as 'instruments of research', to borrow from Longhurst et al. (2008: 215), in conjunction with extended periods of participant observation and immersed engagement with the spaces and displays on offer. We took copious field notes, paying attention to sounds, visuals, movements, narratives and other bodies around us, as well as our own bodily responses (gestures, pulse, responses and emotions). These autoethnographic descriptions were coupled with the collection of still photography, along with audio/video recordings of key displays within the Memorial. What follows is an articulation of the data we gathered in the innovative exhibit dedicated to the Kokoda campaign, which utilizes sound, texture, video and a sense of haunting to focus the attention of the visitor onto a particular narrative. But first we trace the overarching schematic narrative template that is diffused not only through the Memorial but also through wider Australian discourses of nationhood.

Anzac memory: as event and template

To better understand collective remembering, Wertsch (2008a, 2008b) introduces the concepts of 'specific narratives' and 'schematic narrative templates'. As described earlier, the former points to concrete events, people, times or places; the latter refers to more abstract narratives that act as generalized frameworks undergirding specific narratives (Wertsch and Billingsley 2011). The Memorial calls forth a number of specific narratives within its exhibits and displays, such as the Battle of Kapyong during the Korean War, Operation Bribie during the Vietnam War, and a 1942 submarine attack on Sydney Harbour to pluck out just a few. Most, if not all, are underpinned by a deeper narrative that lacks the specificity of concrete events, dates and people but is drawn upon nonetheless by visitors to plot those specific events into a broader script. In the context of the Memorial, we might call this script or template 'the Anzac memory', the archetype of which is undoubtedly the invasion and retreat of the Australian and New Zealand Armed Corps at Gallipoli during the First World War. Commencing on 25 April 1915, the Gallipoli campaign began with an amphibious invasion of the Turkish peninsula, with the ultimate aim of forcing Turkey out of the war by seizing control of the Dardanelles Straits and capturing the city of Constantinople (now Istanbul). The first day of that campaign is perhaps the most documented in Australia's history, remembered as an impossible and bloody task in numerous films, documentaries, books and newspaper articles. In their first major military campaign since Federation, Australian troops landed at Cape Helles and Anzac Cove under the cover of night, emerging onto narrow beaches flanked by steep, deeply incised cliffs alive with enemy fire. Eight months later, with very little gained in terms of military advantage and a humbling number of casualties – some 8,000 lives were lost – the campaign was declared disastrous and all surviving Allied troops were withdrawn (MacDonald 2010).

Clearly Gallipoli presents a specific narrative constructed around particular dates, settings and actions. But in addition to exemplifying a key moment in Australia's military history, elements from that narrative can be abstracted and understood as those that feed into a broader set of sociocultural tools used to author a sense of an Australian identity (West 2008; Donoghue and Tranter 2015). To use Wertsch's terms, abstracted patterns of remembering Gallipoli have become part of the 'general project of developing and maintaining an image that supports a collective identity' (2008a: 68). In other words, parts of the Gallipoli narrative – through repetition over time – operate in ways that allow them to be considered a narrative template. This template can be summarized as one that revolves around a fledgling nation defined by youth, humour, mateship and a sense of adventure that was tested by an ill-conceived and doomed campaign resulting in both loss of life and loss of *innocence* in battle. Through spirit, comradery, endurance and love for country – all now considered central to the Australian psyche – Australia *as a nation* positioned this as a triumph over adversity. In the specific narrative of Gallipoli, this sense of triumph is often underscored with statements that confirm the birth of the Australian nation.

The more generic template has been rehearsed and mobilized countless times since 1915 and continues to permeate Anzac Day ceremonies and other acts of commemoration (see Sumartojo and Stevens, Chapter 11, this volume). Crucially, this framing of memory was immediately and deliberately linked to Australian national identity by Charles E. Bean – a key player in the creation of the Memorial (MacDonald 2010). The Memorial itself captures this form of memory in a display dedicated to 'The ANZAC Spirit today', in which the following quote from William Deane, former Governor-General of Australia (1996–2001), can be found:

> Though born from the doomed campaign at Gallipoli, the spirit of ANZAC is not really about loss at all. It is about courage and endurance, and duty, and love of country, and mateship, and good humour and the survival of a sense of self-worth and decency in the face of dreadful odds.

This, then, is a narrative template with much visibility and *staying power*. And while many aspects of the Gallipoli campaign have remained the subject of debate (the number of lives lost, the tactical ingenuity [or not] of the landing and retreat, prospects for a shared memory with Turkey and so forth), these abstracted qualities remain intact. Tellingly, this is the case despite the increasing diversity of Australia's population over the past hundred years, for whom we might expect to find negotiations of place and belonging of a more transnational nature. Yet support for Anzac memory – witnessed by enduring crowds at Anzac Day dawn services, marches and other ritual events – remains strong; to borrow from Wertsch (2008c: 52), it 'resists change even in the face of . . . social and political transformations'. Undoubtedly, this resistance has been supported politically, finding particular support in former Prime Minister John Howard who did much to revive public interest in Anzac memories during his 11 years in office (MacDonald 2010).

Our focus on Gallipoli as a specific narrative might seem something of a diversion, but it is important for what it reveals about the broader Anzac memory. This narrative template, we argue, sets in place the structure for, and legitimization of, a range of other specific narratives, including narratives around Ned Kelly, the Eureka Rebellion, Changi prisoners of war and the Kokoda Trail Campaign, the latter to which we now turn.

The Kokoda Trail Campaign exhibit

Forming part of the Pacific War of the Second World War, the Kokoda Trail (or Track) Campaign refers to a series of ruthless battles that took place between July and November 1942 in the Australian-administered Territory of Papua on mainland New Guinea. Principally, it was fought by Australian and Japanese troops, though the former were joined by the Papuan infantry battalion and the Royal Papuan Constabulary (Nelson 1997). The trail itself is a tough and rugged single-file jungle track stretching through the Owen

Stanley Ranges, described by Colonel Frank Kingsley Norris in his diary as follows:

> Imagine an area approximately one hundred miles long, crumple and fold this into a series of ridges, rising higher and higher until seven thousand feet is reached, then declining again to three thousand feet; cover this thickly with jungle, short trees and tall trees tangled with great entwining savage vines, through the oppression of this density cut a little native track two or three feet wide, up the ridges, over the spurs, around the gorges, and down across swiftly flowing mountain streams . . . Every now and then leave behind the track dumps of discarded, putrefying food, occasional dead bodies and human foulings . . . In the high ridges, about Myola, drip . . . water day and night softly over the track and through the foetid forest, grotesque with moss and glowing phosphorescent fungi and flickering fireflies. . . .
>
> (Cited in Nelson 2003: 110)

The Campaign consisted of a number of fierce clashes, retreats and withdrawals as Australia attempted to repel a Japanese invasion moving between Gona and Port Moresby along the Owen Stanley Ranges. Commencing with a battle at Awala on 23 July and ending in mid-November, the Campaign was witness to numerous battles in Isurava, Eora Creek, Templeton's Crossing, Efogi, Mission Ridge and Ioribaiwa (Nelson 2003). Like Gallipoli, it is remembered as being conducted on extraordinarily tough terrain that compelled many acts of heroism. It is also remembered as a turning point: not only in military terms with regard to the Pacific War, but for Australia itself, whose fighting image transitioned from 'World War I digger to the green-clad jungle fighter' (Nelson 2003: 125).

The Isurava monument, located in the village of Isurava roughly halfway along the trail, is dedicated to those who fought in the Campaign and commemorates the event with use of the words 'courage', 'endurance', 'mateship' and 'sacrifice' – the schematic narrative template remembered. At the Memorial in Canberra, the precise same qualities are selected for memorialization, fleshed into a narrative that has three core emphases: (1) the individuals that took part in the Campaign; (2) the terribly harsh conditions; and (3) a narration of what Kokoda has come to mean to Australians today – as the place in which the 'soul' of the nation can be found. To supplement this narrative, a new audiovisual presentation and film dedicated to the Kokoda Trail Campaign was introduced into the Memorial in August 2010, as part of the refurbishment of the Second World War galleries. While the presentation is housed within exhibition space that occupies only a small footprint when compared to the sheer size of the Memorial, its importance is nonetheless marked out by virtue of the bounded, dedicated place it is given within what is largely an 'open-plan' gallery. The exhibit itself is located behind a curved wall against which there is a brief history of the campaign, alongside an extract penned by Lieutenant Colonel Frank Norris (a shortened version of the quote cited in this chapter) and an enlarged, original photograph (Figures 10.1 and 10.2). At each end of the curved

Figure 10.1 Inside the Kokoda Trail exhibit.

Source: Authors.

wall is an archway through which the visitor can enter a darkened space adorned with further interpretive panels, soft seating and a large, curved video screen that shows a film playing on a regular loop. From floor to ceiling, the more-or-less enclosed exhibition space is decked out in camouflaged cargo netting, which is loosely hung off the walls to lend the space a textured feeling. There are also life-sized models of soldiers – one standing and the other sitting, almost slumped – placed within a scene depicting the track, filled with tall, straight tree trunks on a steep and muddy incline. This arrangement of space situated the sensing bodies of museumgoers within a particular affective atmosphere; the transition from the openness of the Second World War galleries to the confining spaces of the Kokoda exhibit produced both a vague claustrophobia (with the curving boundaries of the space muddied by the cargo netting and other artefacts) and also an intimate awareness of others' bodies (as potential obstacles restricting movement). While in some ways this sensation mirrors the Gallipoli archetype of the narrative template, with the cramped spaces of the rainforest substituting for the cramped spaces of the trench, it nevertheless also introduces new sensory dimensions: the lush vitality of the greenery, and the verdant potentiality of the mud.

In addition to still photographs, textured adornments and inanimate objects, the Kokoda exhibit also foregrounds artefacts of a very personal nature, including extracts from diaries and letters written to friends and family at home. There is also a video recapitulation of a specific skirmish during the Kokoda campaign,

Figure 10.2 The curved parameter of the exhibit.

Source: Authors.

which splices together photos and first-person narratives undercut with archival footage. This, in combination with the enforced intimacy of the enclosed space, heightened a sense in us of the transpersonal nature of the Kokoda experience (both in 'real life' and as recreated here); it was impossible to engage with Kokoda without also relating to others, both those sharing the space and also the men haunting it. Even when empty of other museumgoers, the space was full – of objects, of stories, of the individual men who fought on the trail.

Indeed, we entered the Kokoda Trail exhibit in a phase we might call 'in-between'; the room was very dark, quiet and empty as the exhibit's film had recently finished and there was an eight-minute delay before the next cycle began. In the interim, the large screen was filled with a static aerial photograph of a jungle village located somewhere along the Owen Stanley Ranges. Black and white in tone, but holding onto that delicate tinge of cool blue that comes with age, the photograph consisted of a small cluster of indigenous-style wooden huts with thatched roofs, surrounded by open, communal space. Small figures were visible in the foreground, frozen mid-stride as they walked bare-chested across the frame or sitting down tending to a task, adorned in pale brimmed hats. Shortly after it was taken, on 27 August 1942, the Australian soldiers figured in the photograph launched an offensive against Japanese forces in the area.

In essence, the exhibit acted for us like a very small, private theatre attempting to conjure something of the harshness of the landscape and the futility of battle. Original photographs and footage of soldiers being pinned down by gunfire, advancing and retreating, were animated with an ominous soundscape and almost monotone narration. Our ability to reach out and touch the cargo netting, the squinting of our eyes as we adjusted to the dark, combined with the overarching auditory overlays brought something *alive* in that space as our eyes returned again and again to those tiny figures. The overall atmosphere had strong echoes with Dylan Trigg's (2012: 82, 98, 79) work on the memory of place, in which he points to the various ways in which materiality interacts not only with the 'making of memories' but the placing of 'our own selves within relation to that past', which in turn invokes the possibility of 'memory being lived as a bodily practice', at least for a moment. Here, we identified a merging of memory with affect, such that the exhibit pivoted upon the use of affect as a means to trigger empathy and position us as having a felt relationship with the 'diggers' depicted. Indeed, as the exhibit reminded us, these young men were trying to triumph, on behalf of the nation, in unfathomable circumstances, with only the leverage offered by a mixture of courage, mateship, endurance and sacrifice.

But rather than understand this narrative through the explicit lens of a nostalgic memory, we point to the role played by our affective potentials and the often intense personal feelings provoked by such exhibits as we attempt to accommodate (or not) a given narrative. This relationship between memory and affect – which we see as a form of embodied remembering – plays out in multiple 'times' and in multiple 'places' simultaneously in the Kokoda exhibit, or elsewhere and else-when, to borrow from Anderson (2014). The Kokoda exhibit poaches a schematic narrative template from visitors' memory-banks and builds itself up with an affectual valence (Byrne 2013) borrowed from another place: Gallipoli. Therefore time and space are compressed between the past and the present, or between historical narratives and shared cultural memories, which in turn pushed to the fore an understanding that is multiple, practised and embodied. As Waterton and Watson (2015) argue, in thinking through these sorts of accounts, we 'reposition "the past" in relation to that of connections made in the present', where 'the former carries with it a meaning that can only be addressed through the feelings we have about our own connections and experiences with the past'. This is a form of bodily remembering we became acutely aware of as we stood in the darkened spaces of the Kokoda exhibit, both before, during and after watching the audio-visual presentation and absorbing the graphic narrative on display. The rugged and steep landscape, coupled with intense battle conditions, seemed to fill with intent and aim to conjure up another memory: our engagements with Kokoda, as audience, seemed to spill out beyond the representations before us and merge with narratives from the broader template. Indeed, our capacities to be affected by memories of Kokoda were qualified by experiences inevitably and already encoded in our person, as well as our responses to already circulating representations. In terms of remembering Kokoda, then, this occurs in concert with a broader narrative, already fleshed out in publicly available collections of signs that include Anzac Day and Anzac memory, and which are mobilized in broader memory projects already steeped in associated, and affirmed, ways of feeling: we are provoked to act in bodily specific ways. Thus, in those moments when a visitor is confronted by small figures of Australian soldiers, moments before battle, their visiting bodies, remembering, are subdued and silenced, rendered 'fit for commemoration' (Allen and Brown 2011: 316). Key to this observation is the idea that neither affect nor remembering is restricted to personal feeling alone – there is a correspondence, or contagion, between bodies and places (see Ahmed 2004).

Conclusions

Our empirical focus on a small exhibit within the Australian War Memorial might seem to overstate the importance of both the exhibit, and the Kokoda Trail Campaign, to Australian collective memory. However, what we have tried to show is how such a small exhibit can contribute to the embodiment of Anzac memory and its associated schematic narrative template. In terms of its narratology, Kokoda can be seen as just another iteration of the template that was set out in the Gallipoli campaign in 1915. Australian national glory cannot be found in any great military success, as in the traditional war narratives of past and present hegemonic powers such as the United States or the United Kingdom. Rather, the Anzac narrative template foregrounds the way in which Australia, fighting alongside these larger powers, triumphs in the face of defeat, pouring national meaning into the notions of valour and sacrifice. Indeed, in these circumstances, Australians are habituated to respond with humour, good will and mateship, as articulated in the Gallipoli archetype of the narrative template. This narrative both allows Australians to hug their allies close and also highlight their distinction from those allies in a way that works to narrate difference. This is a story repeated over and over again in different forms and in specific memorial spaces.

However, it is not enough to narrate such a story iteratively; it must become somaticized in meaningful ways. It is here that the Memorial's Kokoda exhibit becomes important. Its design clearly exceeds the needs of the narrative, most obviously in the deployment of cargo netting and the strict control of light sources to produce a space set apart from the rest of the museum, despite its location right in the middle of the Second World War galleries. Entering this space predisposes bodies to attend to the key sites of narration: the film, the images of Norris and others, and so on. It therefore engenders a sensibility among those participating in the ongoing production of the space; indeed it is the attentiveness of those already in the space, as well as the space's sightlines, which entrain new occupants into opening themselves up to the affective capacities of the media.

Just as the spatial design of the Kokoda exhibit smooths the way for the Kokoda narrative, we also have to highlight the way in which this moment has been temporally engineered as well. For most visitors, this will not be their first exposure to the narrative template of Anzac memory (even overseas visitors are likely to have already seen several incarnations of the narrative in other parts of the Memorial). For Australians, Anzac memory is a commonplace, or perhaps more accurately an all-the-time. Reiterated in various contexts from childhood on, from family histories to school curricula, this schematic narrative template is not just made *legible* but is made *sensible*: 'the history of bodily experience is what sets up a somatic marker profile; in other words, the affective cognition profile of bodies politic is embodied and historical' (Protevi 2011: 402). By bringing Gallipoli, as it is encoded in our bodies, into the Kokoda exhibit we provided a narrative template through which to understand the less-well-known event of Kokoda. Crucially, however, the specific narrative of Kokoda – with its unique geography and historical context – enables the Gallipoli narrative to break free of its specific history, circulating as a set of sensibilities that can be brought to bear on both other aspects of Australian collective memory and also on new 'national' events. This sensibility is not just a lens through which to understand these events, but also a sieve that sorts for what is 'national' in the first place.

Anzac memory is thus rendered into something *felt*: it is comfortable and reassuring. This makes the narrative template itself harder to contest, as it is not merely a story that can be historically disputed; rather, it is a somatic marker that must be undone. 'We have to think of ourselves as bio-historical . . . and with a great deal of plasticity allowing for bio-cultural variance in forming our intuitions' (Protevi 2011: 402). Overcoming such somatic markers for an individual is difficult but possible as a result of the plasticity of the body; overcoming them for a nation is a bigger challenge given the complexity of the political physiology involved.

The diffuse modes through which the narrative template becomes embodied (such as the Kokoda Trail display at the Memorial) can only be countered via a range of media interventions, which work not only to re-narrate Australian identity but also to render the Gallipoli narrative template uncomfortable. By entering into assemblage with the sensing bodies of those encountering these media texts, it becomes possible to unsettle the somatic marker of Gallipoli. This would only ever be an uneven and incomplete enterprise, and by no means would it mean the ascendency of a new narrative. But by unsettling the embodied experience of the narrative, the body is opened up to new sensory experiences, new information and, crucially, a new orientation towards those experiences and information. By understanding collective memory as both more diffuse, and embodied, than has heretofore been accepted, it becomes possible to conceive of a material intervention to re-work that collective memory. The focus of this intervention, however, cannot only be about contesting the narrative; to overturn the narrative template requires an unsettling of the way people *feel* in response to its multifarious narratives.

Note

1 This quote was found on the Australian War Memorial website, available at: www.awm. gov.au/about/.

References

Adey, P. 2008. Airports, mobility and the calculative architecture of affective control. *Geoforum* 39: 438–451.

Ahmed, S. 2004. Collective feelings or, the impressions left by others. *Theory, Culture and Society* 21 (2): 25–42.

Allen, M.J. and Brown, S.D. 2011. Embodiment and living memorials: the affective labour of remembering the 2005 London bombings. *Memory Studies* 4 (3): 312–327.

Anderson, B. 2014. *Encountering Affect: Capacities, Apparatuses, Conditions*. Farnham: Ashgate.

Bartlett, F.C. 1995. *Remembering: A Study in Experimental and Social Psychology*. Cambridge: Cambridge University Press.

Bergson, H. 1988. *Matter and Memory*, trans. N.M. Paul and W.S. Palmer. New York: Zone Books.

Bissell, D. 2010. Placing affective relations: uncertain geographies of pain. In B. Anderson and P. Harrison (eds) *Taking-Place: Non-Representational Theories and Geography*. Aldershot: Ashgate, 79–98.

Byrne, D. 2013. Love and loss in the 1960s. *International Journal of Heritage Studies* 19 (6): 596–609.

Coddington, K.S. 2011. Spectral geographies: haunting and everyday state practices in colonial and present-day Alaska. *Social and Cultural Geography* 12 (7): 743–756.

Connerton, P. 1991. *How Societies Remember*. Cambridge: Cambridge University Press.

Connolly, W. 2002. *Neuropolitics: Thinking/Culture/Speed*. Durham, NC: Duke University Press.

Crang, M. and Tolia-Kelly, D. 2010. Nation, race, and affect: senses and sensibilities at national heritage sites. *Environment and Planning A* 42: 2315–2331.

Crouch, D. 2015. Affect, heritage, feeling. In E. Waterton and S. Watson (eds) *The Palgrave Handbook of Contemporary Heritage Research*. Basingstoke: Palgrave Macmillan, 177–190.

Curti, G.H. 2008. From a wall of bodies to a body of walls: politics of affect: politics of memory: politics of war. *Emotion, Space and Society* 1: 106–118.

Dei, G.J.S. and Kempf, A. (eds) 2006. *Anti-Colonialism and Education: The Politics of Resistance*. Rotterdam: Sense Publishers.

DeLanda, M. 2006. *A New Philosophy of Society: Assemblage Theory and Social Complexity*. New York and London: Continuum.

Deleuze, G. 1968. *Difference and Repetition*. New York: Continuum.

Deleuze, G. 1991. *Bergsonism*. Boston: MIT Press.

Dittmer, J. 2010. *Popular Culture, Geopolitics, and Identity*. Lanham, MD: Rowman and Littlefield.

Dittmer, J. 2013. *Captain America and the Nationalist Superhero: Metaphors, Narratives, and Geopolitics*. Philadelphia: Temple University Press.

Donoghue, J. and Tranter, B. 2015. The Anzacs: military influences on Australian identity. *Journal of Sociology* 51 (3): 449–463.

Dwyer, O.J. and Alderman, D.H. 2008. Memorial landscapes: analytical questions and metaphors. *Geoforum* 73 (3): 165–178.

Fisher, W.R. 1987. *Human Communication as Narration: Toward a Philosophy of Reason, Value and Action*. Columbia: University of South Carolina Press.

Grosz, E. 2005. Bergson, Deleuze and the becoming of unbecoming. *Parallax* 11 (2): 4–13.

Halbwachs, M. 1992. *On Collective Memory*, edited, translated and with an Introduction by L.A. Coser. Chicago: University of Chicago Press.

Hartmann, M. 2015. Why Oklahoma lawmakers voted to ban AP US History. *NYMag*. Available online: http://nymag.com/daily/intelligencer/2015/02/why-oklahoma-lawmakers-want-to-ban-ap-us-history.html, last accessed 15 January 2015.

Heisler, M. 2008a. Challenged histories and collective self-concepts: politics in history, memory and time. *Annals of the American Academy of Political and Social Science* 617 (1): 199–211.

Heisler, M. 2008b. The political currency of the past: history, memory and identity. *Annals of the American Academy of Political and Social Science* 617 (1): 14–24.

Kearney, A. 2009. Homeland emotion: an emotional geography of heritage and homeland. *International Journal of Heritage Studies* 15 (2–3): 209–222.

Longhurst, R. Ho, E. and Johnston, L. 2008. Using 'the body' as an 'instrument of research': kimch'i and pavlova. *Area* 40 (2): 208–217.

Lorimer, H. 2008. Cultural geography: non-representational conditions and concerns, *Progress in Human Geography* 32 (4): 1–9.

McCormack, D.P. 2002. A paper with an interest in rhythm. *Geoforum* 33 (4): 469–485.

McCormack, D.P. 2008. Thinking-spaces for research-creation. *Inflexions* 1 (1). Available online: www.inflexions.org/n1_mccormackhtml html, last accessed 15 January 2015.

MacDonald, M. 2010. 'Lest we forget': the politics of memory and Australian military intervention. *International Political Sociology* 4: 287–302.

Macdonald, S. 2013. *Memorylands: Heritage and Identity in Europe Today*. London: Routledge.

Massumi, B. 2002. *Parables for the Virtual: Movement, Affect, Sensation*. Durham, NC: Duke University Press.

Middleton, D. and Edwards, D. (eds) 1990. *Collective Remembering*. London: Sage Publications.

Muzaini, H. and Yeoh, B.S.A. 2005. War landscapes as 'battlefields' of collective memories: reading the reflections at Bukit Chandu, Singapore. *Cultural Geographies* 12 (3): 345–365.

Nelson, H. 1997. Gallipoli, Kokoda and the making of national identity. *Journal of Australian Studies* 21 (53): 157–169.

Nelson, H. 2003. Kokoda: the track from history to politics. *The Journal of Pacific History* 38 (1): 109–127.

Nora, P. 1989. Between memory and history: le lieux de mémoire. *Representations* 26: 7–24.

Olick, J.K. 2013. *The Politics of Regret: On Collective Memory and Historical Responsibility*. London: Routledge.

Olick, J.K., Vinitzky-Seroussi, V. and Levy, D. (eds) 2011. *The Collective Memory Reader*. Oxford: Oxford University Press.

O'Riley, M. 2007. Postcolonial haunting: anxiety, affect, and the situated encounter, *Postcolonial Text* 3 (4): 1–15.

Orwell, G. 1949. *Nineteen Eighty-Four: A Novel*. London: Secker and Warburg.

Protevi, J. 2009. *Political Affect: Between the Social and the Somatic*. Minneapolis: University of Minnesota Press.

Protevi, J. 2011. Ontology, biology, and history of affect. In L. Bryant, N. Srnicek and G. Harman (eds) *The Speculative Turn: Continental Materialism and Realism*. Melbourne: re.press, 393–405.

Sather-Wagstaff, J. 2011. *Heritage that Hurts: Tourists in the Memoryscapes of September 11*. Walnut Creek, CA: Left Coast Press.

Sellgren, K. 2013. Historians split over Gove's curriculum plans. BBC News. Available online: www.bbc.co.uk/news/education-21600298, last accessed 15 January 2015.

Smith, L. and Waterton, E. 2009. *Heritage, Communities and Archaeology.* London: Duckworth.

Thrift, N. 2008. *Non-Representational Theory: Space/Politics/Affect.* London: Routledge.

Trigg, D. 2012. *The Memory of Place: A Phenomenology of the Uncanny.* Athens: Ohio University Press.

Waterton, E. and Watson, S. 2014. *The Semiotics of Heritage Tourism.* Bristol: Channel View Publications.

Waterton, E. and Watson, S. 2015. A war long forgotten: feeling the past in an English country village. *Angelaki: Journal of the Theoretical Humanities* 20 (3): 98–103.

Waterton, E., Smith, L. and Fouseki, K. 2010. Forgetting to heal: remembering the Abolition Act of 1807. *European Journal of English Studies* 14 (1): 23–36.

Wertsch, J. 2002. *Voices of Collective Remembering.* Cambridge: Cambridge University Press.

Wertsch, J. 2008a. Blank spots in collective memory: a case study of Russia. *Annals of the American Academy of Political and Social Science* 617 (1): 58–71.

Wertsch, J. 2008b. The narrative organization of collective memory. *Ethos* 36 (1): 120–135.

Wertsch, J. 2008c. A clash of deep memories. *Profession*: 46–53.

Wertsch, J. and Billingsley, D.M. 2011. The role of narratives in commemoration: remembering as mediated action. In H. Anheier and Y.R. Isar (eds) *Heritage, Memory and Identity.* London: Sage, 25–38.

Wertsch, J. and Roediger III, H.L. 2008. Collective memory: conceptual foundations and theoretical approaches. *Memory* 16 (3): 318–236.

West, B. 2008. Enchanting pasts: the role of international civil religious pilgrimage in reimagining national collective memory. *Sociological Theory* 26 (3): 258–270.

White, G.M. 2000. Emotional remembering: the pragmatics of national memory. *Ethos* 27 (4): 505–529.

Winter, J. 2006. *Remembering War: The Great War between Memory and History in the 20th Century*, New Haven: Yale University Press.

11 Anzac atmospheres

Shanti Sumartojo and Quentin Stevens

From the early hours of 25 April 2014, people started to gather at the Australian War Memorial (AWM) in Canberra for the annual ritual remembrance of Australia's war dead. 'Australians flock to Anzac Day 2014 Dawn Service', reported the Memorial's press office, with the estimated 37,000 attendees an increase of 2,000 on the previous year (AWM 2014). Meanwhile, in Melbourne, around 60,000 people attended the same event at the Shrine of Remembrance, organised and managed by the Returned Services League and other volunteer groups. Similarly to the Memorial, the Shrine provided a natural backdrop and stage for the speakers, musicians and military personnel in official ceremonial roles. These sites framed and conditioned the experiences of the many people who attended and participated in these events.

One hundred years after the First World War, its commemoration remains a contemporary political concern in Australia and around the world (Sumartojo and Wellings 2014). In Australia, ceremonies such as the Anzac Day Dawn Service provide an example of Winter's 'historical remembrance', a 'discursive field, extending from ritual to cultural work of many different kinds' with a 'capacity to unite people who have no other bonds drawing them together' (Winter 2006: 11). As regular articulations of collective identity, events such as Anzac Day are as personal as they are national, subject to popular and well-attended state rituals that connect participants to the nation.

Such rituals adhere to a narrative familiar to many Australians of 'Anzac nationalism' that uses the historical figure of the First World War solider to personify national characteristics. This story is subject to official control through rules about how the term 'Anzac' can be used and guidelines for how to run an Anzac Day ceremony (Department of Veterans' Affairs 2015). Such ceremonies regularly reinforce this official narrative, with texts, ritual actions and material environments that invite participants to understand themselves and their fellow Australians in particular ways. The places in which these events occur include spatial elements such as symbolic architectural forms, figurative sculpture and lists of 'the fallen' that support Anzac discourse.

As many thousands of people attend and participate in Anzac Day ceremonies in Australia and overseas, their bodies also form part of the ritual. Furthermore, the location of these events at memorial sites both connects commemorants to particular places and to others involved in the same commemoration at the same time in other locations. This chapter focuses on how the participants, material sites and other sensory aspects of the event combine to create potent atmospheres, and how these are both co-constituted by and act upon people who attend these events. The experience of Anzac Day ceremonies at these two memorial sites helps illuminate how the built environment, its activation by national ritual and the sensing bodies of attendees work together to weave together narrative and place into distinctive atmospheres. In turn, these atmospheres help to reinforce particular aspects of Australia's national narratives, and link individuals to much larger collective identities.

Sensory experience and atmosphere

The way the past is evident in, evoked by and reinvented through place is a longstanding subject of enquiry, and memorials provide rich sites to consider this relationship; for example, Treib (2013: xii) remarks that 'architecture and designed landscapes serve as grand mnemonic devices that record and transmit vital aspects of culture and history. Cemeteries and memorials are types of built environments that pursue meaning as part of their making, purposefully.' Memorials can help to define the nation by narrating it symbolically for visitors and residents, but this does not suggest that meaning is settled. Rather, the national identity represented in the built environment is 'not an inert and passive thing, but a field of activity in which past events are selected, reconstructed, maintained, modified and endowed with political meaning' (Said 2000: 185). In their making of national identity, places can thus be conceptualised as 'open' rather than bounded, 'a constellation of processes rather than a thing' (Massey 2005: 141). This constellation includes the many individuals who experience memorials through their own subjectivities, although the 'openness' of place suggests that experiences can unfold that do not rely on particular pre-existing relationships with the represented memorial narratives. Indeed, Wasserman (1998: 43) argues, 'really good memorial places' are those that 'allow the visitor . . . the gaining of experiential insight . . . Through experiencing spaces, viewing and touching artefacts, moving in ritual patterns, and engaging in community activity, the viewer becomes an active participant in the experience of memory.' Here, memorials can help to structure visitors' engagement with the commemorated subject matter through sensory engagement with the sites themselves. However, this is not to diminish the role of visitors themselves in co-producing atmospheres (Edensor and Sumartojo 2015). Instead, we suggest that atmosphere is structured but not limited by spatial environments, and that a range of individual factors contributes to it.

Accordingly, the relationship between place, including memorials, and people who use them has been conceptualised in a range of ways. Thinking about a memorial as a stage, for example, highlights how meaning is made as an active, iterative process, with bodies moving in particular repetitive ways in 'performances like rituals, festivals, pageants, public dramas and civic ceremonies [that] serve as a chief way in which societies remember' (Hoelscher and

Alderman 2004: 350). Böhme (1995) uses a similar metaphor to sketch out 'staged materialities' that use the surrounding matter of placement in the landscape, architectural design and building materials to encourage and enclose the sensory aspects of place. Here, atmosphere emerges as a way of conceptualising the 'feel' of spatial assemblage, mixing together 'narrative and signifying elements and non-narrative and asignifying elements . . . they are impersonal in that they belong to collective situations and yet can be felt as intensely personal' (Anderson 2009: 80). Atmosphere emanates from a dynamic combination of place, people and, in the case of memorials, the traces of national history and individual memory.

Such atmospheres are 'distributed yet palpable, a quality of environmental immersion that registers in and through sensing bodies whilst also remaining diffuse, in the air, ethereal' (McCormack 2008: 413). As participants in atmospheres, it is not clear 'whether we should attribute them to the objects or environments from which they proceed or to the subjects who experience them' (Bille et al. 2014: 2). Indeed, it may be fruitless to try and separate atmosphere from the 'perceiving subject' (Anderson 2009), as it requires the 'sentient subject' in order to emerge (Böhme 2013: 3; Edensor and Sumartojo 2015). Accordingly, individual subjectivity and embodied practices are important parts of how atmospheres emerge, erupt, are sustained and diminished. Meaning is made through interpretation of spatial events and practices, pointing to the implicit role of bodies, with sensory experiences a necessary part of such analytical accounts.

Furthermore, the central figure of the participant means that participation in a commemorative event is shaded with foreknowledge, anticipation, personal memory and, often, family history that produce affective reactions, with the 'obduracy of past experiences that . . . produce a range of feelings' (Rose et al. 2010). McKenna and Ward's (2007: 145) study of Australia's Anzac Day argue that the meanings 'inherent' in the Anzac site at Gallipoli are actually 'inscribed in the assumptions of those who visit the site'. Winter has shown that a majority of visitors to Anzac sites have some foreknowledge, perhaps through a family connection, or some familiarity with Australian military history. She found, for example, that 62 per cent of visitors to Australia's main Western Front memorial at Villers-Bretonneux had a relative who had served in World War I (Winter 2012) and 84 per cent of visitors surveyed at the Shrine of Remembrance were aware of the Anzac landing at Gallipoli (Winter 2009).

If personal history or other previous experiences can shape the experience of commemoration before it even occurs, then visits to memorial sites unveil the role of the senses in making such places meaningful for all visitors. This approach focuses attention on the experiential aspects of memory, activated through the senses and linked to particular sites and spaces, revealing close links among memory, affect and place. For example, in their vivid account of the affective aspects of a visit to the AWM, Waterton and Dittmer (2014: 122) describe in detail their sensory engagement, including the sounds of the eternal flame that 'burbles gently' in the reflecting pool and 'the bugler's pure notes [that] blare out' at the 'Last Post' ceremony marking the Memorial's daily closing time. The actions of the bugle player and the silence that makes the flame audible are part of the assemblage of elements that prompt engagement. This sensory orientation points to the presence of the researcher, engaged in an unfolding process of thinking through his or her own body. Indeed, Carolan (2008: 408) remarks that 'we cannot divorce the mind from the body when talking about knowledge/s, understanding/s and perceptions of the world'. More-than-representational accounts such as this draw on sensory experience, opening up ways of relating place, memory and identity that address 'what momentary experiences look like, and of how meaning (or incoherence) open out into narrative' (Lorimer 2005: 89).

Memorials often provide examples of this process at its most explicit and plain, with emotional and sensory experiences folded into national narratives clearly represented in the material environment. Inscriptions in stone at the Memorial, for example, that praise the 'glorious dead' leave no question of the esteem in which fallen soldiers are held. The solemn and formal physical setting of the Roll of Honour along long alcoves, its bronze plaques listing thousands of names of Australian war dead, encourages contemplative postures and quiet movements. In this chapter, our empirical concern is what the perceiving body makes of memorial environments – thick with national representations – as activated by regular ritual. We ask how this relates to narratives of Anzac nationalism and what a more-than-representational orientation can help us understand about how national identity 'feels' (see Closs Stephens 2015).

Atmosphere and the Dawn Service

Anzac Day, 25 April, commemorates the invasion of the Gallipoli Peninsula by the Australian and New Zealand Army Corps in 1915, a battle that Australians mark as a crucial generative moment of national identity. Indeed, Scates (2006: xxii) remarks that 'Australians discovered their nationhood on the killing fields of Gallipoli'. First commemorated in 1916, its popularity varied throughout the twentieth century, but after a decrease in widespread participation in the 1980s, attendance at Anzac Day events has grown steadily (Holbrook 2014). A century after Gallipoli, 25 April sees Dawn Services across Australia from the official national service at the Memorial in Canberra to small gatherings at local community memorials. Overseas, services occur at many sites, including those in Thailand, Borneo, Turkey, France and the UK. As the public broadcaster, the Australian Broadcasting Corporation covers the Dawn Service live from Villers-Bretonneux, France, the official national memorial on the former Western Front, and Gallipoli, in western Turkey. These ceremonies link Australian commemorative sites across the world through shared activities, broadcast coverage making it 'the commemoration on which the sun never sets' (Bongiorno 2014: 81).

The title 'Dawn Service' explicitly denotes the status of this ritual as a national religious event. The official service in Canberra includes an address from an Army chaplain and the singing of hymns, suggesting an overlap of war remembrance and religious observance. Analogies made in the speeches in Melbourne include references to the Shrine as a 'cathedral of sacrifice' and 'a pantheon of those who have served their country'. Indeed, Seal (2011: 50) argues that Anzac Day 'conflates the sacred and the secular, the military and the civilian with the official and the folkloric in an especially charged moment of time that involves significant numbers of people throughout the country and beyond'. The 'charge' of this event is derived from a combination of site, practices, existing narratives of memory and identity and sensory experience.

To take account of this, our analysis of the Dawn Service unfolds in three streams: feelings of anticipation; spatial design and symbolic representations, including the bodies of 'enactors'; and immersive sensory experiences of the event. We explore how these experiential aspects of the ceremonies link to narrative elements, and how this might connect individual attendees to national identity. Indeed, a sense of national connection is implied by the very act of attending an Anzac Day ceremony, given its status as a national holiday commemorating the deaths of military members who died in the service of the state (or in the case of the First World War, the Empire). The ubiquity of national identity in structuring how we understand society and everyday life (Billig 1995), political assembly (Closs Stephens 2013) and popular culture (Edensor 2002) make it a compelling lens for examining collective experience.

In terms of methodology, we concentrate on our observations of the eventful environment, including how spaces were organised, what happened in them, how we observed others responding and also how we reacted, using our own bodies as 'instruments of research' (Longhurst et al. 2008). We conducted research on Anzac Day 2014, Stevens attending the Dawn Service at the Shrine of Remembrance in Melbourne, the most highly attended state-level service, and Sumartojo the same event at the Memorial in Canberra, the main official national service, which is broadcast live across Australia from 5.30am. We used auto-ethnographic methods, attending to our own experiences of the event, supplementing our field notes with video and audio recordings and still photographs that captured our experience of the Dawn Service and the sensory aspects of the environment we encountered. We also analysed print and broadcast media, focusing on descriptions of the sites and attendees' bodily movements, postures and actions within them. We were interested in the Dawn Service because it is commonly reported as emotional and atmospheric, and is also collective, political, consciously designed, ritualised and presented as representative of a national identity. This combination of factors suggested the possibility of distinctive 'Anzac atmospheres'.

Setting the scene: anticipation, performance and staged bodies

In studies that range from shopping centres to festivals to heritage sites, preparation for and anticipation of events has been shown to condition experience and shape atmospheres (Rose et al. 2010; Edensor 2012). At the Dawn Service, this was readily observable in how people were dressed, with most prepared for the chilly pre-dawn conditions. As researchers, our own rituals of preparing clothes and bags for the next day, setting alarms, going to bed early, and otherwise planning for an early start showed our anticipation of the service. We each observed many people gathering or approaching the ceremonial sites in small groups, often multi-generational families, making phone contact with others, or meeting up. This suggested a level of forethought or organisation before the event, and also the involvement of other people in personal commemoration. Media accounts suggest that many have some idea of what they will experience before they go (Marshall 2014). Certainly the popular narrative of Anzac is widely known to most Australians, even those without a family connection, learned through schooling, widespread media coverage, and popular cultural expressions such Anzac Day football matches or the sale of Anzac biscuits (Scates 2002; Holbrook 2014).

At both sites, anticipation was also built into the physical approach to the site. In Canberra, although the researcher could hear the amplified sound of people giving speeches at the Memorial, the site was only revealed slowly as she walked closer. The traffic and parked cars on the approach became more concentrated the nearer she got, and finally the extent of the site became fully visible as she found a place to stand, feeling hurried to get a 'good' spot by the many other people entering the site. Turning the corner into the site, brightly projected images from the Memorial's archival collections of military personnel and battle place names were visible on the main façade of the building, a practice that had begun on Anzac Day 2013. This heightened the impression that something significant was happening.

In Melbourne, the urban design around the Shrine also helped to increase anticipation. The long northern axial approach along Swanston Walk and St Kilda Road contained many people walking quietly toward the Shrine from the city centre, with the crowd thickening close to the site. Several spatial cues heightened the sense of choreographed anticipation: the long, slow incline; the incremental widening of the paved pathway; its tight perspectival framing by dark rows of cypresses and the dramatic temporary up-lighting of the Shrine's colonnaded portico. The processional way was punctuated by a short, steep wide stair-case leading onto the Shrine's forecourt, which slowed the crowd's approach, and by the wide paved cross-axis of the Second World War forecourt with its flanking eternal flame, cenotaph and flagpoles, symbolic reminders of national commemoration.

This hints at how the locations and material representations in the memorial buildings themselves contributed to a sense of purpose and solemnity. In Canberra, the Memorial, completed in 1941, sits at the foot of Mt Ainslie, terminating the axis that extends across Lake Burley Griffin from Parliament House. It was designed in 1928 in a Byzantine Art Deco style with a central copper-clad dome and a symmetrical façade punctuated by a flight of stairs between two long, blocky wings. There is little external decorative detail (RSTCA 1986). The stated functions of the Memorial are to 'commemorate the sacrifice of those Australians who have died in war . . . [and] to assist Australians to remember, interpret and understand the Australian experience of war and its enduring impact on Australian society' (AWM 2015a). The sombre sandstone building plays an important role as a backdrop to regular national rituals. Extending down a gentle hill from the central entrance are three shallow flights of stairs that lead to a Stone of

Figure 11.1 The pyramidal form of the Shrine of Remembrance with the crowd gathered in its forecourt.

Source: Photo by Quentin Stevens.

Remembrance, a simple altar-like tablet that is a repeated element at Australian First World War grave sites. An important designer of Commonwealth cemeteries, Edwin Lutyens, intended this stone to be a 'universal architectural expression of an imperishable mass' as well as 'an altar as a place of religious actions and the offer of sacrifices' (Geurst 2010: 22). Thus, although it was not intended to be an explicitly religious symbol, its form strongly hints at Christian notions of sacrifice and an afterlife.

Melbourne's Shrine of Remembrance is modelled on the tomb of King Mausolus in Harlicarnassus, in eastern Turkey, an amalgam of the archetypal funerary pyramid, tower, and temple to a god (Figure 11.1). The Shrine publicises its classical antecedents, describing the relief sculptures on the northern tympanum as 'a winged Goddess, symbolic of Mother Country, calling her children to defend her' (Shrine of Remembrance 2014). The Shrine is widely visible from its position on a high hill terminating the axis of one of Melbourne's major commercial streets, and the uphill procession to it on Anzac Day reinforces the gravity and importance of the events it commemorates. The monumental and 'exotic' architecture of both memorials hints at Gallipoli's Turkish location and reinforces the sombre and religious qualities of the Dawn Service.

In 2014, attendees at the Dawn Service in Canberra sat in seats in rows arranged at the front of the Memorial and in raked up bleachers behind the rows of seats. Others stood on the red-gravel surface of the building's forecourt and on the sloping lawns flanking the central courtyard. The participants were oriented towards a podium behind the Stone of Remembrance that served as a spatial and symbolic focus for the event. The spatial organisation of the crowd was very different in Melbourne, where people stood in a mass before and below the Shrine, facing it; from the perspective of the crowd, the hill and the building's tapered profile exaggerated its scale. In their assembled stillness, attendees at the ceremony effectively became part of the extended architecture of the memorial. Many people stood listening with heads slightly bowed, the flat forecourt making it difficult to see the speakers on the Shrine through the crowd.

During the Service, speakers, singers and military personnel were framed in front of the Shrine, highlighted against its dark mass. The building's axial symmetry, its pitched roof and pediment and the contrast of its dark-shadowed porch against the light-coloured stonework all drew the onlookers' gaze to the performers' central location; they appeared as both representatives and guardians of the site and its meanings. This was reinforced by the rifle salute directed up and away from the Shrine over the crowd, and the official party's privileged first entry inside the building at the end of the ceremony, after which members of the public were invited to enter the Sanctuary and pay their respects, under the guidance of marshals.

The Dawn Service thus takes place in designed environments where architecture, including the approach to the site, spatial organisation, the control of the movements of attendees, and the choreographed actions of officials and participating military personnel all play their part. However, as we will discuss next, sensory elements of the setting also play an important role in the atmosphere that coalesces around the Dawn Service.

Immersed bodies: sound, light and temperature in Anzac atmospheres

Rather than a passive 'audience' for a display or performance, attendees at the Dawn Service are participants, co-constituting the atmosphere of the event through their own memories, sensory perceptions, reactions and anticipatory urges. The surges and flows of atmosphere that cling to these events imbue them with a powerful affective charge evident in common descriptions of them as 'bloody amazing' or 'really moving' (McKenna and Ward 2007). But they also have sensory impact that mixes intimately with the ceremonies' emotional and narrative aspects. Thus, even as attendees move through or observe these elements, as Böhme (2013: 4–5) reminds us,

> [t]he spaces generated by light and sound are no longer something perceived at a distance, but something within which one is enclosed . . . It becomes clear that what is at issue is not really visual spectacles . . . but the creation of 'tuned' spaces, that is to say, atmospheres.

Light, sound and temperature were three elements that helped 'tune' immersive Anzac atmospheres during the Dawn Service. The feel of the event was strongly shaped by the pre-dawn conditions, and the interplay of light and darkness, both artificial and natural, worked to set this event apart from everyday events. In Melbourne, the Shrine precinct was very dark after the lights of the street during the approach. The researcher there experienced it as feeling confined and dark, defined by nearby light sources. It was only as dawn approached that the silhouettes of the surrounding trees became noticeable, making the setting feel less intimate as it was revealed as a large, open space under the brightening sky. Architectural lighting was used to aesthetic affect and to draw attention to particular symbolic elements of the building. Although the Shrine's eternal flame remained lit in the Memorial's forecourt, attendees' attention was directed forward and upward, particularly through bright illumination. Before the Service, the portico was uplighted, but as the formal prelude began, it was darkened and 'The Symbol of Glory' atop the Shrine's pyramidal roof was illuminated. Both sets of lighting drew the gaze of participants and created spatial drama and focus, and the latter had an ethereal, mysterious dimension that amplified the Service's religious character.

In Canberra, in was so dark that as the service began the researcher had only a poor sense of how big the area was where she was standing, or even the size of the crowd. Her attention was drawn to the silhouettes of trees and surrounding buildings, and the colour of the sky as it changed and lightened. The approach to the Memorial was well-lit by streetlights and passing cars, but the grounds were much darker. The central Stone of Remembrance was also uplighted and a big screen cast a flickering glow over the crowd. The dark conditions highlighted the bright, striking images of Australian service people and battle place names projected onto the two pillar-shaped wings at the front of the Memorial building (Figure 11.2). The glow and sharp definition of the projections, framed by an imposing architectural form, blended the narrative and sensory aspects of the event (Barns and Sumartojo 2015).

The transition from darkness to gentle dawn illumination meant the crowds were shadowy and indistinct at both sites until the sun rose. The researchers' awareness of being one individual among many was heightened as the extent of the crowds became apparent. The dawn scheduling of the ceremony has its own connection to the Anzac narrative, recalling the timing of the attack at Gallipoli:

> [I]n the setting of the Dawn Service, dawn's temporality is poignant because it suggests the inevitability of changed circumstances: the Anzac mythology associates dawn with action, movement, noise and the likelihood of violent death. But it also brings a type of relief, and end to the awful waiting before battle . . . At the Dawn Service, darkness retells and reinforces the narrative of the Anzac attack at Gallipoli, working this into the bodies of commemorants through their sympathetic experience of dimness, stillness and anticipation.
>
> (Sumartojo 2015)

This recalls Ingold's (2000: 258) description of light as 'a phenomenon of experience, of that very involvement in the world that is a necessary precondition

Figure 11.2 The crowd at the Dawn Service at the AWM in 2013, with the projections
visible on the façade of the main building.

Source: Photo by Quentin Stevens.

for the isolation of the perceiver as a subject'. Thus, the artificial illumination that
highlighted particular symbolic elements of the memorial sites helped enrol
attendees in an Anzac narrative, while the transition from night to dawning light
shaped symbolic and sensory aspects of the ceremony.

Sound – music, voice, silence and the various noises of birds, wind, traffic and
bodies moving – also enfolded participants, immersing them in a solemn, still
and contemplative atmosphere. Marshall (2004: 40) remarks that since the end of
the First World War,

> state-sanctioned remembrance has been characterized by a specific . . . series
> of sounds used at every scale [that] . . . create[s] the ritual's sonic boundaries
> and delimit[s] its territory just as the symbolic space surrounding a war
> memorial forms a material boundary.

Hymns, bugle music, the rhythm of speeches and repetition of ritual texts all
worked to signal ritual practices of remembrance during the Dawn Service.
Approaching the Memorial during the 'pre-show' before the official 5.30am start
of the national live broadcast, the researcher could hear the amplified voices of
readings from diaries and letters, before the site or the speaker were visible. This
included an address from Wing Commander Sharon Brown, a Royal Australian
Air Force medical officer who had served in Afghanistan. Brown's account of her

war service and her feelings of sadness, pride and empathy encouraged respectful attention, combining an intimate reflection on war experience with the impact of a single amplified voice in an otherwise quiet gathering: 'As the dawn delivers the daylight, pause to reflect upon the memory of those that have gone before us in your name, those whose faces and names grace the walls of our Memorial' (Brown 2014).

A new addition to the 2014 service was a didgeridoo performance, played by serving Navy member Able Seaman Darren Davies. The low, deep sound of the traditional instrument rolled across the shadowy crowd, its vibrations seeming to reverberate in the researcher's body, a very powerful sensory experience. This recalls Ong's (1982: 82) remark that hearing allows us to 'gather sound from every direction at once: I am at the centre of my auditory world, which envelops me, establishing me at a kind of core of sensation and existence . . . You can immerse yourself in hearing, in sound.' This was the first time that the indigenous instrument had featured at the national service, and knowledge of the history of Aboriginal inequality made it a political experience as much as a sensory one. This was heightened by an ongoing debate about the location and content of an Aboriginal war memorial. At present, there is no official national memorial specific to Aboriginal and Torres Strait Islander service people, despite demand for one, although there have been recent installations in state capitals Sydney and Adelaide (Armbruster 2014). There is also an unofficial one on the slopes of Mt Ainslie behind the main Memorial, built by a private benefactor that hosts a commemorative ceremony following the official Dawn Service (Bongiorno 2014).

Another notable aspect of the Dawn Service was silence, both as a traditional feature of the ritual and as a powerful generator of a commemorative atmosphere. Marshall (2004: 41) reminds us that the 'interplay of sound and silence within public remembrance is central in transforming the everyday landscape in which the war memorials are located and ceremonies take place into places set apart from the quotidian'. Elements such as the music of bugling and singing were expected noises for national military remembrance, and the traditional two-minute silence was not unusual. However, the silence was remarkable because of the large crowd in attendance. For both researchers, it was noticeable: the usual buzz and murmur of thousands of other people simply was not there. This was enhanced by the crepuscular conditions that meant the extent of the crowd was unknown. Any unexpected sounds were quite shocking. For example, in Melbourne, some attendees arrived on thunderous motorcycles at their designated parking area near the Shrine, their loudness rousing the drowsy pedestrians. At the same service, loud salvos of rifle shots punctuated the stanzas of the hymn 'Abide with Me'. While this was startling, their reverberations helped define the size of the space, otherwise difficult to discern in the dark. Sonic elements such as rifle shots, the sound of the bugle and the specific military refrains – the 'Last Post' and the 'Reveille' – also reinforced the Anzac narrative.

The collective incantation 'Lest We Forget' encouraged attendees to identify sonically with the crowd, and by extension that nation, and to pledge to remember the Anzac dead. The sound and the collective activity of hymn singing supported an atmosphere that enveloped the audience, enhancing a sense of common purpose. Collective speech, however, 'suggests unity but actually conceals diversity' (Marshall 2004: 41). As researchers, our reactions to the sounds of the services varied. In Melbourne, for example, the loud arrival of motorcycle riders before the service reminded the researcher of the fear-inspiring 'Ride of the Valkyries' played from helicopters in the film *Apocalypse Now,* and thus of the Vietnam War that some of the motorcyclists may have fought in. In Canberra, the hymn 'Abide with Me' had a powerful and unexpected emotional effect in its invocation of comfort and solace at death. Thus, as researchers we were not passive receptacles for overwhelming or pervasive atmospheres, but co-constituted the atmosphere with our own bodies, memories and reactions, as immersed and engaged 'perceiving subjects' (Anderson 2009; Edensor and Sumartojo 2015).

The final element of our account is the temperature and weather conditions that we experienced at the Dawn Service, taking up Ingold's (2005: 102) characterisation of weather as a 'medium of perception'. The embodied experience of the very early start and the chilly autumn temperatures in Melbourne and Canberra are strongly linked to the narrative of Anzac sacrifice. In his Dawn Service address, Corporal Ben Roberts-Smith began by linking his 'overwhelming sense of pride' in Australia with 'the cold and dark of another Anzac Day dawning' (Roberts-Smith 2014). The Memorial, in its material on the Anzac Day arrangements, informs visitors the early start is a reminder of the dawn landing on Gallipoli in 1915 (AWM 2015b). The walk to the Shrine and patient waiting in the tree-framed space in a large crowd on a dark, cold morning allow attendees to imagine the uncertainty, suffering and sacrifice of the journey and arrival in Gallipoli. The Dawn Service gestures towards that moment through the bodies of attendees, encouraging remembrance through an analogous bodily experience. In their willingness to stand still and silent in generally uncomfortable conditions, members of the audience affirm and are reminded of their respect for the fallen. As this report from the Melbourne newspaper *The Age* shows, that is part of how people understand their experience:

> The ceremony is held 'in a chilling half-light that is not yet morning and no longer night . . . This was the time of day when battle came . . . and one had to be alert and ready and attentive . . . 'it's a mark of sacrifice,' said Mr Dorling. 'We owe that much.'

(Marshall 2014)

When we attended Anzac Day ceremonies, although the morning was cold, at both sites this was more noticeable after the ceremony than before and during it. The large crowds were very still and quiet, but the closeness and warmth generated a distinctive micro-climate that wove together the thermal and the emotional. Here, we experienced a sense of togetherness that reinforced collective identity and sentiment, complemented by focused points of light and of the speakers' and singers' voices. With the attenuation of other sensations in the quiet, cold, still darkness, the amplified voices of the orators seemed to reach out and into our bodies.

Anzac atmospheres: memory and lived experience

As Marshall (2004: 38) insists, 'we use our senses to forge connections with our physical environment and develop our sense of place . . . Remembrance, because it is experienced through the senses, is one such embodied state.' In attending Anzac Day ceremonies at special memorial sites, we joined thousands of other people enacting remembrance with and through their bodies. Indeed, many recent memorials afford tactile, haptic, kinaesthetic and sonorous experiences that potentially 'touch' people and 'resonate with' their understandings of the past. Through the 'mnemonics of the body', a visit to a memorial site often provides 'a knowledge and a remembering in the hands and in the body . . . it is our body – at a subconscious level, rather than our mind – which "understands"' (Connerton 1989: 95; Stevens and Franck 2015). In addition to memorials' visual appearance, we found that commemoration was made meaningful through embodied, multi-sensory experiences.

Accordingly, at the Memorial and the Shrine of Remembrance, bodily experiences of light, sound and temperature helped provoke engagements and responses from visitors that transmitted meaning and reinforced narrative. The Dawn Service engendered perceptual experiences and bodily engagements where interpretations of Anzac were acted and felt as well as read. The bodily aspects of remembering (Connerton 1989) were both evident and important, as were a range of collective ceremonial actions that affirmed memorials' commemorative and symbolic purposes.

In co-constituting the atmosphere of the Dawn Service through their participation in ritual activities, including with their movements, voices and subjective interpretations, attendees at the ceremonies contributed to the narrative that the memorial sites conveyed, extending Connerton's argument that memory is 'sedimented' in the practices of the body (1989: 72). Dawn Service attendees participated in remembrance and contributed to its meaning through their own actions and postures. The combination of bodily and cognitive engagement with the rituals of the Dawn Service can have an intense impact on attendees, triggering affective responses that make a visit meaningful. The metaphoric and mnemonic potential of such spatial experiences is vivid in the reactions that draw together physical and affective experiences, as in the repeated descriptions of Anzac Day as 'moving' (Thompson 2014; McKenna and Ward 2007).

The collective nature of these experiences also helps to connect participants to other contemporary Australians through participation in a crowded site where the postures, noises, bodily warmth and reactions of other people contributed to the atmosphere of the event. Attendees were also connected to Australian co-nationals elsewhere who were participating in the same rituals in similarly early conditions. Thus, in the case of the Dawn Service, national identity is 'felt' at a range of scales: individual bodily sensations; local collective experience; and national co-ordination of commemorative ritual. In its repeated reference to a century-old military campaign, the national community on Anzac Day is also imagined as 'deep, horizontal comradeship' (Anderson 1991: 7) that reaches far into the past.

Accordingly, on Anzac Day Australian memorials reverberate with atmospheric intensities generated by the combination of the subjective – personal connection, foreknowledge of narrative, embodied sensory experience – and the objective. The Dawn Service involves the careful organisation of memorial spaces to accommodate attendees, the repetition of music, hymns and well-known ritual texts, and the display of national symbols. Together, these elements are taken up and enriched by attendees, alongside additional aspects – illumination and darkness, noise and silence, weather and temperature, the presence of a crowd – to create potent atmospheres of collective identity. This resonates with the claim that 'the staging of atmosphere is a way of being together, of sharing a social reality' (Bille et al. 2014: 4). The familiar contours of the Anzac narrative – that Gallipoli saw Australians distinguish themselves both as soldiers and as distinctly national – work to engender feelings of common purpose that the spatial and experiential elements of the event reinforce. This is not to argue that these are experienced as uniform. Receptivity to particular atmospheres, as Duff remarks (2010: 881), 'is largely an expression of the social ties that form its foundation', based in part on previous personal experience. However, as researchers, our own experiences of the Dawn Service were of distinctive Anzac atmospheres, crafted by a range of both representational and sensory elements, taken up and animated by our fellow attendees.

Acknowledgements

This research was supported by an Australian Research Council Future Fellowship (project number FT0992254) and a grant from the RMIT University School of Architecture and Design Research Committee. Thanks also to Felicity Cull for her invaluable assistance with the research.

References

Anderson, B. 1991. *Imagined Communities: Reflections on the Origin and Spread of Nationalism*. London: Verso.
Anderson, B. 2009. Affective atmospheres. *Emotion, Space and Society* 2: 77–81.
Armbruster, S. 2014. Veterans push PM for Indigenous war memorial. *SBS News*, 7 July. Available online: www.sbs.com.au/news/article/2014/07/07/veterans-push-pm-indigenous-war-memorial, last accessed 13 April 2015.
AWM. 2014. Australians flock to Anzac Day 2014 Dawn Service. Press release, 25 April. Available online: www.awm.gov.au/media/releases/australians-flock-anzac-day-2014-dawn-service/, last accessed 27 January 2016.
AWM. 2015a. About the Australian War Memorial. Available online: www.awm.gov.au/about/, last accessed 19 February 2015.

AWM. 2015b. Dawn Service. Available online: www.awm.gov.au/commemoration/anzac/dawn/, last accessed 23 February 2015.

Barns, S. and Sumartojo, S. 2015. When one idea led to another: re-inscribing and recombining thinking spaces using night-time projections at the Australian National University. *The Senses and Society* 10 (2): 179–199.

Bille, M., Bjerregaard, P. and Sorensen, T. 2014. Staging atmospheres: materiality, culture, and the texture of the in-between. *Emotion, Space and Society* Available online: http://dx.doi.org/10.1016/j.emospa.2014.11.002, last accessed 18 January 2016

Billig, M. 1995. *Banal Nationalism*. London: Sage.

Böhme, G. 1995. Staged materiality. *Daidalos* 56: 36–43.

Böhme G. 2013. The art of the stage set as a paradigm for an aesthetics of atmospheres. *Ambiances*. Available online: http://ambiances.revues.org/315, last accessed 23 February 2015.

Bongiorno, F. 2014. Anzac and the politics of inclusion. In S. Sumartojo and B. Wellings (eds) *Nation, Memory, and Great War Commemoration: Mobilizing the Past in Europe, Australia and New Zealand*. Oxford: Peter Lang, 81–98.

Brown, S. 2014. Anzac Day pre-Dawn Service address 2014. Available online. www.awm.gov.au/sharon-bown-anzac-day-2014/ [*sic*]. last accessed 24 February 2015.

Carolan, M. 2008. More-than-representational knowledge/s of the countryside: how we think as bodies. *Sociologia Ruralis* 48 (4): 408–422.

Closs Stephens, A. 2013. *The Persistence of Nationalism: From Imagined Communities to Urban Encounters*. London: Routledge.

Closs Stephens, A. 2015. The affective atmospheres of nationalism. *Cultural Geographies*. Available online: http://cgj.sagepub.com/content/early/2015/02/17/1474474015569994. abstract, last accessed 27 January 2016.

Connerton, P. 1989. *How Societies Remember*. Cambridge: Cambridge University Press.

Department of Veterans' Affairs. 2015. Protecting the word Anzac. Available online: www.dva.gov.au/commemorations-memorials-and-war-graves/protecting-word-anzac, last accessed 23 February 2015.

Duff, C. 2010. On the role of affect and practice in the production of place. *Environment and Planning D: Society and Space* 28: 881–895.

Edensor, T. 2002. *National Identity, Popular Culture and Everyday Life*. Oxford: Berg.

Edensor, T. 2012. Illuminated atmospheres: anticipating and reproducing the flow of affective experience in Blackpool. *Environment and Planning D: Society and Space* 30: 1103–1122.

Edensor, T. and Sumartojo, S. 2015. Designed atmospheres: introduction to special issue. *Visual Communication* 14 (2). 251–266.

Geurst, J. 2010. *Cemeteries of the Great War by Sir Edwin Lutyens*. Rotterdam: 010 Publishers.

Hoelscher, A. and Alderman, D. 2004. Memory and place: geographies of a critical relationship. *Social and Cultural Geography* 5 (3): 347–355.

Holbrook, C. 2014. *Anzac: An Unauthorized Biography*. Sydney: New South.

Ingold, T. 2000. *The Perception of the Environment*. Abingdon: Routledge.

Ingold, T. 2005. The eye of the storm: visual perception and the weather. *Visual Studies* 20 (2): 97–104.

Longhurst, R., Ho, E. and Johnston, L. 2008. Using 'the body' as an 'instrument of research': Kimch'i and pavlova. *Area* 40 (2): 208–217.

Lorimer, H. 2005. Cultural geography: the busyness of being 'more-than-representational'. *Progress in Human Geography* 29 (1): 83–94.

McCormack, D. 2008. Engineering affective atmospheres on the moving geographies of the 1897 Andrée expedition. *Cultural Geographies* 15: 413–430.

McKenna, M. and Ward, S. 2007. 'It was really moving, mate': The Gallipoli pilgrimage and sentimental nationalism in Australia. *Australian Historical Studies* 38: 141–151.

Marshall, D. 2004. Making sense of remembrance. *Social & Cultural Geography* 5 (1): 37–54.

Marshall, K. 2014. Large crowd turns out for Melbourne dawn service. *The Age*, 25 April. Available online: www.theage.com.au/victoria/large-crowd-turns-out-for-melbourne-dawn-service-20140425-zqz5h.html, last accessed 18 February 2015.

Massey, D. 2005. *For Space*. London: Sage.

Ong, W. 1982. *Orality and Literacy: The Technologizing of the Word*. London: Methuen.

Roberts-Smith, B. 2014. 'Anzac Day Dawn Service Address', AWM. Available online: www.awm.gov.au/anzac-day-dawn-service-address-2014/, last accessed 23 February 2015.

Rose, G., Degen, M. and Basdas, B. 2010. More on 'big things': building events and feelings. *Transactions of the Institute of British Geographers* 35 (3): 334–349.

RSTCA (Register of Significant Twentieth Century Architecture). 1986. Australian War Memorial. Available online: www.architecture.com.au/docs/default-source/act-notable-buildings/r016_australian_war_memorial_rstca.pdf?sfvrsn=2, last accessed 19 February 2015.

Said, E. 2000. Invention, memory and place. *Critical Inquiry* 26 (2): 175–192.

Scates, B. 2002. In Gallipoli's shadow: pilgrimage, memory, mourning and the Great War. *Australian Historical Studies* 33 (119): 1–21.

Scates, B. 2006. *Return to Gallipoli: Walking the Battlefields of the Great War.* Cambridge: Cambridge University Press.

Seal, G. 2011. '. . . and in the morning . . .': adapting and adopting the Dawn Service. *Journal of Australian Studies* 35 (1): 49–63.

Shrine of Remembrance. 2014. The Shrine story: self-guided tour of the Shrine of Remembrance, brochure. Available online: www.shrine.org.au/Shrine/Files/55/55ad2875-440c-40ee-9c81-fab9c9ade2ae.pdf, last accessed 25 February 2014.

Stevens, Q. and Franck, K. 2015. *Memorials as Spaces of Engagement: Design, Use and Meaning*. New York: Routledge.

Sumartojo, S. 2015. On atmosphere and darkness at Australia's Anzac Day Dawn Service. *Visual Communication* 14 (2): 267–288.

Sumartojo, S. and Wellings, B. (eds) 2014. *Nation, Memory, and Great War Commemoration: Mobilizing the Past in Europe, Australia and New Zealand*. Oxford: Peter Lang.

Thompson, A. 2014. 'Austinmer Anzac Day Dawn Service 2014: photos'. 25 April. *Illawarra Mercury*. Available online: www.illawarramercury.com.au/story/2240501/austinmer-anzac-day-dawn-service-2014-photos/, last accessed 23 February 2015.

Treib, M. 2013. 'Yes, now I remember': an introduction. In M. Treib (ed.) *Spatial Recall: Memory in Architecture and Landscape*. Abingdon: Routledge, x–xv.

Wasserman, J. 1998. To trace the shifting sands: community, ritual and the memorial landscape. *Landscape Journal* 17 (1): 42–61.

Waterton, E. and Dittmer, J. 2014. The museum as assemblage: bringing forth affect at the Australian War Memorial. *Museum Management and Curatorship* 29 (2): 122–139.

Winter, C. 2009. The Shrine of Remembrance Melbourne: a short study of visitors' experiences. *International Journal of Tourism Research* 11: 553–565.

Winter, C. 2012. Commemoration of the Great War on the Somme: exploring personal connections. *Journal of Tourism and Cultural Change* 10 (3): 248–263.

Winter, J. 2006. *Remembering War: The Great War between Memory and History in the Twentieth Century*. New Haven: Yale University Press.

12 Beyond sentimentality and glorification

Using a history of emotions to deal with the horror of war

Andrea Witcomb

The defence of ANZAC Day commemoration – as common in the 1920s as today – turns on some fairly familiar arguments. It does not glorify war; it does not cultivate hatred; it is about honouring and remembering, not celebrating. Yet a sense of sacred nationhood created through the blood sacrifice of young men remains at its core today, as in 1916. Is this not to glorify war?

(Bongiorno 2015)

Working at the Australian War Memorial many years ago I became aware that the Memorial, from its inception, had deliberately, and perhaps properly, avoided much engagement with the emotions of the museum visitor. Many war museums and interpretation centres, even in recent times, have gone down exactly the same path. *Love & Sorrow* entirely rejects this approach. This is an exhibition that openly and deliberately works on the emotions of its visitors to proclaim its strong and powerful message: war is an unmitigated and abhorrent disaster and we need always to be conscious of its enduring impacts across subsequent generations. This exhibition is anything but a celebration of the centenary of the First World War.

(McKernan 2015)

These two quotations, both in response to commemorative activities around the First World War in Australia during the 2015 centenary celebrations, seem to pull in contrary directions, opening up a space for productive reflection. The first, by Frank Bongiorno (2015), comes from a piece in which he traces the history of the discursive qualities of the Anzac (Australian and New Zealand Armed Corps) mythology that underpins the Australian commemoration of war, pointing out that the emotive language around sacrifice, bravery and bloodshed is not only designed to elevate the ordinary individual into a hero and celebrate his capacity for 'mateship' but also to prevent any critical engagement either with the history of war or the way Australians have commemorated it. The result is not only to celebrate war but it is also to produce Australia's war experience as a sacred one, making any form of critique profane. In effect, Bongiorno is suggesting that this celebratory language is emotive and that it produces an uncritical nationalism.

McKernan's (2015) comment in his review of *WWI: Love & Sorrow* at the Melbourne Museum that the majority of exhibitions dealing with war are not concerned with producing emotional responses in their visitors seems jarring, or at the very least to point in a contrary direction. Referring to his own former institution, the Australian War Memorial, McKernan paints a depiction of its exhibition practices as straight representations of history when read in contrast with the overt emotionality of *WWI: Love & Sorrow*. While not disagreeing with his point that this overt emotionality is part of the exhibition's strength, the assumption that the exhibitions at the Memorial and elsewhere are not emotive, I want to argue, point to the need for a more complex understanding of the possible relations between history and emotion as well as the ways in which emotions are embodied within exhibitions.

In this chapter, then, I explore the relationship between history and emotion in two different kinds of exhibitions on war – one based on military history, the other on social history. This is in order to posit a distinction between two different kinds of history-making in exhibitions about war in Australia. The first kind, which is prevalent at exhibitions within a memorial setting such as a Shrine of Remembrance or the Australian War Memorial, is emotional history – a history that produces an uncritical sentimentality that results in feelings of pathos. The second kind is a history of emotions, a genre of history-making that, following the work of Theodore Zeldin (1971, 1982), Barbara Rosenwein (2002) and William Reddy (2001), posits that emotions are central to understanding human experience and offer a unit of analysis for historians wishing to understand not only the past but also its legacy in the present. While both forms of history offer emotional experiences, they do so to different ends and using different forms of affect. My second aim is to undertake an analysis of the ways in which affect is produced across these two different approaches to the use of emotions and to ask what spaces for 'critical pedagogy' (Witcomb 2013) these strategies open or close. I will prosecute these analyses by comparing two new exhibitions commemorating the First World War in Melbourne, Australia. The first is at Melbourne's Shrine of Remembrance, where there is a new suite of exhibitions commemorating Australia's wartime experiences, including the First World War, and the second is *WWI: Love & Sorrow* at the Melbourne Museum. My methodology is essentially auto-ethnographic in that I explore my own sensorial and emotional engagement with these exhibitions. I do so first of all by spending time in these exhibitions, taking them in. I then photograph individual displays and take video recordings of multimedia installations as well as making notes of particular moments that arrested my attention in order to document the exhibition in ways that help me access particular details later such as the use of language, particular images, juxtapositions, layout and use of colour for example. I look for patterns and sequences that help to anchor and explain my experiences at the exhibition and then seek to analyse how these elements work together to create that experience. At that point I look for theoretical insights that help me to describe what is going on both in terms of structure and its effects.

Before beginning that analysis, however, I want to sketch the history of debates in Australia over how the First World War has been remembered and commemorated, as these set the frame for the different functions of each type of exhibition. In the immediate aftermath of the First World War, the association between war and nation-making was very strong, due in large part to the legacy of an Edwardian belief that war purified the nation and clarified its values, testing the mettle of its men (Reynolds 2010). While Australia's experiences at Gallipoli were an unmitigated disaster, it was the moral character of the soldiers that shone through in the ways that that disaster was recuperated for public consumption, giving rise to the association between war and Australian national identity. This allowed any need to question Australia's uncritical support for the Empire to simply disappear from public view.

However, as Donaldson and Lake (2010) argue, there was also a strong narrative that the best way to commemorate and honour the dead was to prosecute an argument for peace between nations, rather than to continue to validate the morality and worthiness of war as the crucible in which nations are formed. As they put it, '[t]he view propagated by Anzac mythology that World War I was a creative experience for Australia in that it made us a nation was an obscene idea for many in the 1920s and 1930s' (Donaldson and Lake 2010: 73). Many wondered what the enormous loss had been for, realizing that the association between war and nationhood only brought misery; some felt that Australians should not have been in the war at all and that we were fools to answer the call of Empire so readily; others found that their status as Anzac heroes was not matched by the lack of opportunity they had in post-war society. For others, honouring the dead only allowed a forgetting of those who survived. As Joan Beaumont has argued, Australia was a broken society in the aftermath of the First World War (Beaumont 2013) and there was no unity around how best to commemorate the Anzacs. The pacifist strain of thought re-emerged during the Vietnam War, with this period also seeing extensive criticism of the Anzac legend around the masculinity of its narrative, its erasure of the act of killing, as well as its racist and imperial foundations.

Contrary to expectations of the demise of the Anzac narrative, however, in the 1980s, a new discourse around the figure of the Anzac emerged, which produced the ordinary soldier as a victim who rose above his historical circumstances. He became, as Donaldson and Lake (2010: 90–91) put it, a 'tragic hero', one who died too young, still innocent of the ways of the world – and one with whom it was all too easy for Australia's youth to empathize. Rather than generating anger, as the narrative of heroism had in the immediate aftermath of the war, historians such as Joy Damousi (2010) argue that this new narrative generated both sentimentality and nostalgia amongst the young for their forebears. Placing these particular forms of emotion at the heart of commemoration, she argues, enables the erasure of any need to understand either the history of the events themselves or of their commemoration. As she puts it '[a] critical examination of the costs and consequences of war, its horror and waste, the mistakes and massacres is resisted and repressed' (Damousi 2010: 97). In other words, raising questions about either agency or responsibility is understood as a form of profanity. The consequence is that there is no room to question the claim that Australia's national identity is not only founded in its war experience but also in the qualities now promoted as characteristic of all Anzacs and first forged in the crucible of Gallipoli. As a number of concerned historians have pointed out (Lake et al. 2010; Luckins 2004), it is as if the work of building a strong democratic society is not an essential part of what it is to be Australian, with the effect that alternative narratives around equity and social inclusion are simply erased as an important part of our history.

The Melbourne Shrine of Remembrance

The problems are in many ways illustrated by a new suite of exhibitions at Melbourne's Shrine of Remembrance dealing with all the major wars in which Australians have been involved as part of a significant extension to their exhibition galleries. Opening in time for the centenary commemorations of the First World War at the end of 2014, these exhibitions are chronologically arranged, moving from one war to another but with no attention to the period between the wars and very little on the home front. Such a focus on military history rather than the history of war results, as Beaumont (2013) has argued in relation to the First World War, in the lack of an understanding of the impact of the war on Australia. In many ways, this is a conventional history – empirically based, with plenty of documentary evidence, making it the kind of exhibition McKernan is referencing in his introduction to his review of *WWI: Love & Sorrow* quoted at the beginning of this chapter. Even its design supports its empiricist epistemology. Minimalist in its aesthetic, objects are hung or placed in exhibition cases in neat linear rows. Everything is rigidly straight, plain and restrained, as is the floor plan – no nooks, intimate spaces or quiet corners. Rather like a three-dimensional book, intent on providing information about the course of the war, there is no particular use of design to create a sense of drama, of different scales and experiences. There is no special lighting, hardly any ambient music or use of sound – except, that is, for one significant display that provides the emotional heart of the exhibition and that reinforces the feeling that, despite all the force of a chronological approach based on original documents, we are in a sacred space that works by pulling at our emotional relationship with the Anzac legend. How does this happen? How is the stage set for this one display to have such emotive force?

There are three key strategies that enable an affective connection to the deep collective forms of attachment to the Anzac narrative in ways that privilege an emotional attachment rather than a rational appreciation of the historical information presented by the makers of the exhibition. These are the use of particular forms of language, the aesthetic ambience of the space itself and, lastly, the use of the human face and voice, all of which are set within a space that is a shrine, with all the connotations that that carries.

Language

An emotional rather than rational relationship to the Anzac tradition and the histories to which it refers is set outside the exhibition galleries, as one enters the interpretation centre through a memorial-like atrium on whose walls are engraved the following words: 'ANZAC is not merely about loss. It is about courage and endurance, and duty, and love of country, and mateship and good humour and the survival of a sense of self-worth and decency in the face of dreadful odds.' Apart from the use of the present tense, which binds the readers into the message, these are the sacred words that Bongiorno (2015) alludes to as constituting the basis upon which we celebrate war as the foundation of the Australian nation. These words embody the coming together of the mateship narrative that historians on the left developed during the 1950s to symbolize national identity (see Ward 1958 for example) with the narrative of sacrifice and courage at the centre of the Anzac legend. As well as identifying the moral qualities of those we are commemorating and seeking to emulate, these words also point to the emotional qualities of the narrative itself, which are just as important to the Australian sense of self they seek to identify. For these qualities emerge as the result of story that is told as a tragic epic, a narrative form that promotes awe as well as sadness or pathos. We are meant to identify with these men and feel what they felt.

The first attempt to actively shape our emotional orientation to the exhibition once inside the building is a short visual installation that summarizes the function of the Shrine and its exhibitions. Once again, language is the key to the way this installation works, serving to link what one is about to see – mainly a 'straight' history of the various wars in which Australia has participated from our point of view – to our current ways of framing the Anzac legend for nationalistic purposes. The first sentence focuses on the 'stories of the men and women honoured within these walls' for their 'service and sacrifice'. The focus on individuals is immediately apparent, as is the narrative of sacrifice. Both of these provide the means through which the history of war is made central to the nation. Central to this narrative is a focus on those 'who did not return' and in whose memory the Shrine was 'built by those who grieved for them', with its bricks 'laid by hand by those who'd been to war'. The Shrine, therefore, is a sacred place of memory that embodies the grief of those who sought to remember the dead forever. The ongoing remembrance of those who died, then, is central to the work of the Shrine and its visitors. Peppered throughout the displays, therefore, are little vignettes detailing the lives of those who served, though the narrative itself is not structured by their experiences but by the relentless chronology of the war itself. Those we now remember, however, also include the builders of the Shrine – the family members who grieved, the mates who returned; for theirs was a sacred responsibility that we, according to the Anzac narrative, have now inherited.

Aesthetic ambience

Just as important to communicating the sacred nature of the Anzac heritage are the aesthetic qualities of the space itself. Situated within the foundation of the Shrine, the exhibition material is in part organized in a linear manner because it is set within the spaces created between the rows of brick foundation columns. We can feel that we are underneath the Shrine with its eternal flame and the space itself is reminiscent of a church with its rows of columns supporting the vaulted roof above and leading inexorably to the altar that symbolically contains the body of Christ. The exhibition too has its altar, albeit in a side chapel rather than at the end of the nave.

Created by the removal of one of the brick pillars supporting the Shrine above, this 'chapel' houses a lifeboat from the *SS Devanha*, a former merchant navy ship turned troop carrier that took troops to Anzac Cove on the Gallipoli Peninsula in Turkey, in what proved to be a military disaster for the Allied Forces but the beginning of the Anzac legend for Australians. At a quarter to and past the hour, a sound, image and light show plays over the boat, creating the emotional heart of this gallery. Set at dawn, washed by moonlight, the audience is transported back to the moment of the first landing at Anzac Cove. We wait with the soldiers in the dark, hearing only the sounds of oars gently splashing the waves. A voiceover using first-person snippets from those who 'were there' encourages us to imagine their fears as they glide silently across the water. The tension is increased by our knowledge that the Turks are waiting too, having seen the silhouette of the transport ships moving silently towards the shore. Everything is still, and yet all the senses are alert, watching, waiting. 'I watched and waited,' says a Turkish soldier, slowly and deliberately. We too watch and wait, aping his behaviour. We watch above all the faces of the men portrayed in the black and white photographs taken at the time and projected onto four vertical screens behind the boat. What were they thinking and feeling as they sat in motionless silence in those boats slowly taking them to the Turkish shore?

Faces and voices – the creation of a powerful assemblage

Faces are important, Anna Gibbs (2010) tell us in her essay on the ways in which affect spreads amongst humans like a contagious virus. Following the work of psychologist Sylvan Tomkins, who identified nine different affects, stretching from negative to positive ends of a continuum such as, for example, pain to pleasure, Gibbs points out that it is the face that registers these affects the most and that communicates it to others. The face and the voice. Thus, for example, if someone smiles it is very hard not to smile back, not only offering a recognition of their smile but also as a result of feeling the joy that original smile gave us. Smiling back provides an intensification of that affect. It is something that happens automatically, a visceral response from one body to the other. For Tomkins,

> affects are not private obscure intestinal responses but facial responses that communicate and motivate at once publicly outward to the other and backward and inward to the one who smiles or cries or frowns or sneers or otherwise expresses his affects.
>
> (Tomkins 1966: vii, cited in Gibbs 2010: 191)

Metaphorically sitting side by side with these men on the lifeboat from the *SS Devanha*, it is very hard not to get caught up in the affects and emotions being registered by these men on their faces, feeling their apprehension, their fears, all the more because we also know the end of the story. Many would not come back. As one primary school girl softly whispered to her friend at the end of the show, 'that is so sad'.

But I am jumping ahead. For our apprehension is acknowledged through the use of the human voice as well, as what these men were waiting for suddenly becomes clear as Australian voices, in a range of accents representing our connections back to Britain as well as the emergence of a clearly definable Australian accent, provide descriptions of the scene taken from contemporary accounts. 'The first thing we heard was a single shot. Then two or three, then it began very fast,' says a clearly Australian voice. Another comments that 'we were completely helpless and exposed in our slow moving boat'; 'just target practice,' says a working-class voice in a slightly cockney accent. 'We had hoped to take the enemy by surprise,' says an educated officer in a clearly discernable Australian accent while the foot soldier repeats, 'just target practice'. In those few lines, communicating the recorded thoughts and memories of a range of Australian soldiers across the ranks who took part in the landing at Gallipoli, is embodied what little history is known by Australians reared by the representations in Australian cinema and television since the late 1970s of the Gallipoli story. It goes like this: the Australians were regarded as cannon fodder who could be dispensed with by their English superiors who bungled the whole thing anyway. The men's lives were wasted but their personal and collective spirit as Australians triumphed (see also Dittmer and Waterton, Chapter 10, this volume). A voiceover provided by a stiff-upper-lipped Australian officer steeped within the ideology of Empire and with an English accent to match confirms that interpretation, as if giving assent to that belief from the very centre of Empire: 'It was a magnificent spectacle, to see those thousands of men rushing through the hail of death as though it was some big game,' while a laconic Australian voice simply says, 'Here we are, now we are in it,' as the sound of the waves wash over the audience and presumably the men in those boats. Pride in their courage is tangible as is their faith in the Edwardian belief that war made nations. As the Turks fire back, making it impossible for the Anzacs to advance much over the difficult terrain, a proud officer comments: 'These chaps don't seem to know what fear is. A barge will be sufficient to take us back.'

The point of all this oblique representation of death was not whether or not there was any reason as to why Australians should have been involved in the first place. There are no grand arguments made here as to the righteousness of their cause. The point is simple, as the voice of an upper-class Australian officer in an English accent tells us: 'The Australians did heroic work and the world will know it.' The point is to celebrate the military courage of these men, a point that is made sharper by the emotional pathos created by our knowledge that so many did not return. The men are rendered victims, without agency despite their bravery. And our agency, or rather our responsibility, is only to mourn them and not ask questions. The point is made clear in the last voiceover, in an extract from a letter written from a soldier to his wife at home in the knowledge that he could die, requesting her and their daughter, even commanding them – and in this context, us – to 'never forget'. After the sound and light show, photographic panels with the faces and names of soldiers who died at Gallipoli on that first landing are shown, one after another. Their youth is clear for all to see but it is their courage and not the waste of young lives that we commemorate. The engagement is based on an emotional identification with them; there is no space for a critical political enquiry into the events.

Attempts to resurrect a moment in history and to place us there, particularly those associated with difficult and traumatic histories, have a long history of critique, and for good reason. They share in the problem that Dominick LaCapra (2001) identified in relation to the use of victim testimonials at the site of the original trauma in films such as Claude Lanzmann's *Shoah*. In wanting to produce an empathic relationship between the victim and the film audience – such films were, LaCapra argued, in danger of replaying the original trauma by not having sufficient distance from the moment of origin. In placing the victim back at the site of their original trauma, lending their testimony the authority of the place where those events took place, time is made to stand still. By erasing temporal distance, audiences were given no outside from which to engage with the reenactment. While LaCapra's discomfort was out of concern for the victim, who was left in a continual state of victimhood, LaCapra also argued that it was naive to assume that it was possible to take on the victim's position. As an audience, we know the trauma is momentary, he argued, that it does not really belong to us. To pretend that we can step into someone else's shoes is, for LaCapra, to wallow in dangerous sentimentality. Rather than calling for the need for empathy, then, LaCapra suggests that what we need are strategies that, in their interpretation of difficult histories, cause an 'empathic unsettlement' that enables us to ask questions out of concern for the victims but that nevertheless returns to them their agency over their own destiny. To be able to do that, we need to be able to connect their experience to that of our own, in our own time. We need to normalize the subjects of the past.

Normalizing the Anzacs is not, however, what this display achieves. Instead, the representation of Australia's wartime experiences as one in which ordinary men become heroes is in the tradition of monumental history. Monumental history, as Friederich Nietzsche (1997) tells us, is concerned with providing moral lessons – examples of best behaviour that we can emulate in the present. While in Nietzsche's time such examples came from politicians and military figures, the Australian innovation, is, as Graeme Davison (2000) points out, to elevate the ordinary man to such a position, particularly the foot soldier. Hence, the language of Anzac frames both the Shrine and the exhibitions within it as a commemorative act. This should not be a surprise, given the memorial functions of this site, but it does remind us that the power of the place and the language that informs it is such that no attempt at a positivist form of historical narrative is going to overcome its function in myth-making for purposes of nation building.

WWI: *Love & Sorrow* at the Melbourne Museum

WWI: Love & Sorrow offers a contrast to the Shrine, making no attempt to explain the causes and course of the war. Its aim is neither military history nor monumental history. It has none of the traditional language associated with the commemoration of the Anzacs. Blood, sacrifice, courage, mateship and bravery are neither concepts nor words that can be found anywhere within the exhibition. There are no examples of ordinary men and women as heroes, though individuals are the focus of its narrative structure. Rather than being represented as an epic story, war is the setting for the destruction of families and communities in ways that echo down through later generations. Tragedy, rather than the epic, is its narrative form.

This narrative form is given shape by a focus on eight individual and family biographies through an analysis of two of their emotions as they develop in response to the experience of war – love and sorrow. In tracing and giving body to these two emotions, curator Deb Tout-Smith has used them to provide not only a heart-wrenching story as to what happened to each and every individual and their family but also to provide a window into the effects of the war on Australian society more generally. The exhibition is a rendition of the way in which war breaks society by breaking its emotional heart – the home. It is thus no surprise that the exhibition contains a representation of a fireplace at the centre of its geography, signifying the hearth. Such a centre allows for a consideration of the ways in which society itself was broken as the impact of the war impinged on the home front – read both as the nation, the local community and the family, across all classes and racial and ethnic divides.

Visitors to the exhibition are invited to follow the intensity of people's love for one another as well as manifestations of their pain when things go wrong. These intensely private experiences are difficult to witness and all the more so as by the end we come to realize that the pain they embody is still felt two or three generations later – a theme that has its echoes with Marianne Hirsch's (2001) discussion of post-memory within families that survived the Holocaust. It is this pain that opens up a range of political questions that are suggested in snippets of contextual information that run alongside the eight biographies. They range from the question of what it was all for, to the duty of care by the state to those who survived, to awkward questions concerning unequal treatment around racial divides, to the longevity of the social and economic impact of the war, to the weight born by families of those who returned with their spirit or their bodies broken.

How affect is deployed as an interpretative strategy to achieve this is what I want to focus my analysis of this exhibition on. In essence, my argument is that these strategies are responsible for enabling a form of 'empathic unsettlement' (LaCapra 2001), precisely because the exhibition refuses to maintain a linear narrative between past and present. Instead, it builds connections across time and between people. It does so both within the families represented and those who end up as witnesses to what is, in effect, their testimonies via a form of mimetic communication (Gibbs 2010) that builds on the ways in which the exhibition produces 'sticky objects' (Ahmed 2010: 29) that embody the emotions of love and sorrow between people, enabling a contagion-like spread of these between generations and to the audience. In my interpretation of how the exhibition works, affect is a central medium of communication and emotions are both the subject of enquiry and the aim. Together with the help of the information that is provided within the narrative aspects of the exhibition, this provides the ground for a form of critical enquiry on the part of the visitor.

What then is mimetic communication and how does it work? In her essay, Gibbs describes mimesis as a form of communication practice that embodies relations between people, rather than the communication of information. For Gibbs, this involves

> corporeally based forms of imitation, both voluntary and involuntary . . . At their most primitive, these involve the visceral level of affect contagion, the 'synchrony of facial expressions, vocalisations, postures and movements with those of another person', producing a tendency for those involved 'to converge emotionally'.
>
> (Hatfield et al. 1994: 5, cited in Gibbs 2010: 186)

As explained above, in the discussion on faces, there are a range of affects such as, for example, joy, anger, pain, disdain and pleasure that are communicated through facial expressions (grimaces, smiles) and in the tonal quality of the human voice (loud, quiet, soft, harsh, fast, slow, rising, diminishing), which, as argued by Tomkins (1966: vii, cited in Gibbs 2010: 191), spread contagion-like, across bodies. As the surface traces of deeply felt bodily sensations, these affects help to build emotional landscapes that build as well as break social bonds (Gibbs 2010: 191). Ultimately these affects contribute to building a sense of belonging to or being excluded from society.

Within the exhibition, this form of communication is built up through a particular assemblage of objects, images and voices, built into a narrative sequence in which time is at once chronological, in that we follow the lives of particular people, but is also flat in that the temporal distance between those who experienced the events that are the subject of the discussion and their descendants and visitors to the exhibition is broken down and we are encouraged to feel the same emotions they did via this process of mimesis. The difference between this exhibition and the one at the Shrine is that this process of mimesis is not aimed at getting us to be in their shoes (the resurrection of the past) but to ask questions about the larger narrative in which this historical event is usually framed within Australian society.

Of particular importance to the power of this mimetic form of communication is the way in which a small group of objects become 'sticky objects'. This is a concept developed by Sara Ahmed (2010) to describe the ways in which objects can accrue layers of meaning that become attached to them and which leave a residue on those who come into contact with them. This residue is what she describes as affect. For her, affect 'is what sticks, or what sustains or preserves the connection between ideas, values and objects' (Ahmed 2010: 29). In doing so, she is also subscribing to the view that the material world has agency, that it has particular forms of power that can draw people as well as other objects into a field of relationality with them. Understanding these relations is to understand how affect works and what it produces.

Mimetic communication in WWI: Love & Sorrow

While each of the eight biographies we can follow in the exhibition have a 'sticky object' and a witness to their stickiness, here I only have space to focus on three examples. We first come across the eight individuals we are invited to follow across the exhibition through black and white portrait photographs of them. These are set against other photographs of them with their families, as if on a mantelpiece, suggested by a small narrow shelf on which their portrait is placed underneath the larger family shot. Succinct text underneath these photos provides enough biographical detail to enable us to situate them as part of a wider family, their ethnicity or race, their age, where they lived and what they did for a living. The theme of love and the suggestion of possible sorrow are introduced with a short extract from a letter written by a soldier to his wife that says: 'If I am to die, know that I died loving you'. Amongst them is Frank Roberts, then 27 years old, who married his fiancée, Ruby, months before he left. His parents, though worried, were proud that he was doing his 'duty'. Albert Kemp was a butcher who lived in Normanby Street, Caufield with his young wife Annie and their two small children, one of them a newly born baby at the time he left. John Hargraves was an 18-year-old from Horsham in Victoria, a small country town, who worked as a telegraph boy. He had to have the permission of his parents to enlist as he was so young. The bonds between these three individuals and their families were strong, a point that is made through an assemblage of objects that includes photographs, letters, telegrams and personal effects that have been kept within these families and are still owned by them. Some of these are used a number of times, becoming 'stickier' each time we meet them. Thus, for example, in the panel that tells us fairly early on in the exhibition that Frank Roberts was sent to France, we have a black and white photograph of him looking up at his sweetheart, Ruby, adoringly. She looks down on him as she is lifted up above him with a gentle smile, placing her arms around his neck in an embrace. It is obvious that they adore one another. The negative space between their faces is charged with love and that is what our eyes focus on, as this is the centre of the image. The figure of their baby daughter, born in November 1917, the same month that Frank was sent to the trenches in Belgium, is referred to in his diary, where he wrote that as he lay in his trench waiting for the orders to go over, 'the picture I had before my eyes was that enlarged photo of you and little Nancy'. The next time we come across this family is to learn of Frank's death in Belgium. We do so with that enlarged photograph of Ruby and baby Nancy before us. Beside it is a little package containing a letter purportedly from Nancy to her father with one of her little booties asking him if he would be able to fit into it. The letter and the little bootie never made it to Frank, and were returned to Ruby. She kept them all her life, later giving them to Nancy. At the end of the exhibition, in a movingly filmed interview with Ruby's own granddaughter, the little bootie turns up again, this time in Ruby's granddaughter's hands. Unconsciously, as she talks of Ruby's pain and the hole that Frank's death had on their family, she strokes the little bootie, as if to console her mother and grandmother. Sticky with the residue of the pain experienced by this family, including that of Frank's father who spent the rest of his life compiling every bit of information about Frank and his life in a series of scrapbooks in a labour of love that also documents his own grief, we become witnesses to the ways in which this war had a lasting effect on Frank's family and his descendants. As we watch Ruby's granddaughter fight back her tears and notice her involuntary caresses of the little bootie, we cannot but respond mimetically, metaphorically caressing her back with our own affective and emotional response. We become part of the circle of sorrow.

Albert Kemp is one individual whose biography is central to the exhibition, forming part of its emotional core. He wasn't particularly brave but he was loved and he loved in return. Admitting to his own fears in a letter to his wife Annie, we feel for his little daughter Ethel, then five years old. She looks out at us a number of times through the repeated use of an image of her taken from a family portrait produced to send to Albert. She seems a vulnerable little girl, looking at us with wide, open eyes and a very serious demeanour. In the first showcase introducing us to the family, there is a postcard she had written to her dad in which she tells him 'Dear Daddy I am waiting and watching every day for you.' Immediately round the corner from this, we hear the only sounds of guns in the entire exhibition. Walking into this sonic landscape I found myself standing between a Memorial Roll and the landscape of Glencourse Wood – a desolate piece of no man's land, ripped apart by cannons. As I walked through the space I encountered myself in a direct relationship to the past, for the display involves the slow dissolve between three images. Two are historical, taken by the English and the Germans documenting the progressive destruction of the woods. The third is a present-day image of the same place – the woods have returned, the birds are singing, everything is fresh and green. Life has returned, but it is on top of bodies that were never recovered. 'Tread softly by/Our hearts are here/With our beloved Jack', says the text behind, quoting the words put on the grave of Jack Reynolds by his parents. The request means something, for as we move through the space the silhouettes of our bodies cause the dissolution from one image to another. We are present in that landscape and provide continuity between past and present. The act of walking is an act of remembrance. Walking out we are faced almost immediately with the image of little Ethel's postcard and her face again – this time beside a reproduction of a telegram sent to the local priest asking him to communicate Albert Kemp's death at Glencourse Wood to his wife Annie. His body was never found and I had just tread softly on his place of death.

Around the corner we come to an Edwardian fireplace that is not only the hearth of every Australian home but also the hearth of Albert Kemp's family. On the mantelpiece is a Memorial Plaque given to his family. Beside it, the family portrait from which the image of little Ethel we saw as we took in the news of his death is taken. To the side is a photograph of Frank Roberts, festooned in mourning ribbons and placed in the family's dining room. Around these objects are many other mementoes from other familiar hearths. To the left of this display is a multimedia interactive detailing the impact of war on all the residents of Normanby Street, where the Kemp family had lived before the war. Together with the hearth, it is clear that Ethel's story and that of her mother, of Frank, Ruby and Nancy is repeated not only on that street but also throughout the land as Albert and his family become a microcosm of a global experience let alone a national one. At the end, Ethel's god-daughter, a woman now in her middle years of life holds Ethel's postcard to her dad as she explains the impact of his death on his family. Her thumb caresses it, as if she was offering comfort to Ethel and Annie, acknowledging not only their sorrow but also physically, viscerally, embodying her own witnessing of this sorrow. In turn, in reading her face, her voice and her gesture, we too become witnesses and ponder on not only the emotional impact but also what it meant in practical terms – having to leave the family home, live with in-laws and deal with economic hardship.

John Hargraves provides the basis for the most heart-wrenching critique of the war. A young man who set off for adventure as an 18-year-old, he came back suffering post-traumatic stress as a result of gassing on the front as well as being buried alive. While he recovered to some extent, his trauma was always with him. At the beginning of the exhibition, we sense his excitement on going to the front. The possessions kept by his family include his runner's armband, a photograph of him with two of his 'mates' also in uniform and his war diary. It is this last object that becomes 'sticky' in the hands of his daughter Joan as she reads to us from it and then, looking straight at the camera, tells of the impact of his trauma on himself and the family. She herself is clearly traumatized by it but she is not a victim. While tears well up in her eyes and she covers her face in her hands as the camera dissolves away, this is not before she delivers her stinging critique based on her experience of the effect of the war on her father. 'It gave me a horror of wars,' she says, going on to conclude 'it is just incomprehensible why we continue to go to wars', indicating in this one statement what the purpose of the exhibition was – to support a pacifist argument by clearly indicating the ongoing traumatic, social and economic costs of war.

In many ways, this exhibition is an example of what Theodore Zeldin describes as a history of emotions based on individual biography.[1] In an article about the nature of social history, Zeldin (1971: 242) advanced an argument as to the need for historians to be freed of what he called the tyranny of chronology and a conception of time as being able to be periodized as if the past bore no relation to the present. For him it was important to 'show that the past is in fact alive' rather than to 'resurrect' it (Zeldin 1971: 242). In his own work, Zeldin seeks to do this by concentrating his unit of analysis on individuals and their biography, rather than on abstract categories such as class, gender, race and ethnicity, nation or community – or, in our case, Anzacs. He is interested in how people's emotions play a part in shaping their approach to the world around them and thus a role in defining what opportunities they have to act in the world. Understanding the histories of these emotions is thus one of the keys to explaining past events (history) – a point that is equally relevant to those of us who study the construction of heritage and are concerned about the ways in which narratives of heritage can confine one's horizons and ability to engage across difference. It is important, he argues, for historians to 'liberate themselves from the frameworks they inherit from their predecessors' (1971: 243), just as it is for all people to recognize that the narratives they use to understand themselves are historically produced and therefore can be changed. The future, he suggests, can only be different, if we understand the history to our present. He is aware that doing so makes it impossible to establish causality, as the more one looks at individual biographies, the more complex people's agency becomes. Nevertheless, he maintains that a focus on the individual does produce an understanding of society at any given time and across time.

Significantly, given Zeldin's comments on the importance of letting go of chronology and his resistance to the notion of resurrecting the past, the assemblage of objects, images and voices at work in this exhibition work, as we have seen, to destabilize any notion of the past as disconnected from the present while emotions are given life by the ways in which a particular suite of objects weave in and out of the story at particular moments in time, building the emotional intensity as well as an understanding of the significance of the two emotions under scrutiny – love and sorrow – to the messages this exhibition embodies. The past, in a real sense, collides with the present and with us, the witnesses to other people's pain. Not simply mementoes, these 'sticky objects' become 'sites of feeling' (De Nardi 2014), which in turn have the potential to become what I have elsewhere called a 'pedagogy of feeling' – a form of cultural pedagogy that uses affect to open up a critical engagement with received narratives about the past (see Witcomb 2015a, 2015b).

The effectiveness of this strategy as well as the difficulties it faces in countering the weight of the Anzac legend can be seen in the written responses people leave behind on a comments wall designed to enable people to respond to the exhibition and the emotions it elicits. While the great majority of responses gesture towards the 'lest we forget', there are a significant number that indicate support for the pacifist argument as well as pass comment on current foreign policy. Examples of statements made by a wide variety of people include questions such as 'Why are we still involving ourselves in war?' from a child; 'It was a war we didn't have to fight in. So many died and yet we still celebrate'; or 'I look forward to the day when war does not occur, when we do not live in fear with greed and power dominating our selfishness preoccupations, when we have the courage to vote in leaders who are honest, egalitarian and will not allow killing in the name of our country.' One who gave new meaning to the phrase 'lest we forget', embodied what I would argue is the aim of this exhibition – to question what the dominant language around Anzac hides and in the process to produce a position from which we can be more engaged citizens: 'May those who have gone before give us pause to remember and reflect before we fight the wars of tomorrow. Lest we forget what we have lost.' This, it seems to me, encapsulates the critical potential of forms of remembrance that use a history of emotions to refuse a distinction between past and present and recover the voices of those we have forgotten in order to ask questions about the present.

Note

1 With many thanks to Julian Thomas for suggesting Theodore Zeldin to me after a joint visit to the exhibition with our families.

References

Ahmed, S. 2010. Happy objects. In M. Greg and G.J. Seigworth (eds) *The Affect Theory Reader*. Durham, NC: Duke University Press, 29–51.

Beaumont, J. 2013. *Broken Nation: Australians in the Great War*. Sydney: Allen & Unwin.

Bongiorno, F. 2015. A legend with class: labour and ANZAC. *The Conversation*, 21 April 2015. Available online: https://theconversation.com/a-legend-with-class-labour-and-anzac-38592, last accessed 30 April 2015.

Damousi, J. 2010. Why do we get so emotional about Anzac? In M. Lake, H. Reynolds with M. McKenna and J. Damousi (eds) *What's Wrong with Anzac? The Militarization of Australian History*. Sydney: New South/UNSW Press, 94–109.

Davison, G. 2000. *The Use and Abuse of Australian History*. Sydney: Allen & Unwin.

De Nardi, S. 2014. An embodied approach to Second World War storytelling mementoes: probing beyond the archival into the corporeality of memories of resistance. *Journal of Material Culture* 19 (4): 443–464.

Donaldson, C. and Lake, M. 2010. Whatever happened to the anti-war movement? In M. Lake, H. Reynolds with M. McKenna and J. Damousi (eds) *What's Wrong with Anzac? The Militarization of Australian History*. Sydney: New South/UNSW Press, 71–93.

Gibbs, A. 2010. After affect: sympathy, synchrony, and mimetic communication. In M. Greg and G.J. Seigworth (eds) *The Affect Theory Reader.* Durham, NC: Duke University Press, 186–205.

Hirsch, M. 2001. Surviving images: Holocaust photographs and the work of postmemory. *The Yale Journal of Criticism* 14 (1): 5–38.

LaCapra, D. 2001. *Writing History, Writing Trauma*. Baltimore and London: John Hopkins University Press.

Lake, M., Reynolds, H. with McKenna, M. and Damousi, J. 2010 (eds) *What's Wrong with Anzac? The Militarization of Australian History.* Sydney: New South/UNSW Press.

Luckins, T, 2004. *The Gates of Memory: Australian People's Experiences and Memories of Loss and the Great War*. Fremantle: Curtin University Books/Fremantle Arts Centre Press.

McKernan, M. 2015. Review of *WWI: Love & Sorrow*. *ReCollections* 10 (1). http://recollections.nma.gov.au/issues/volume_10_number_1/exhibition_reviews, last accessed 30 April 2015.

Nietzsche, F. 1997 [1873–1876]. *Untimely Meditations*. Edited by D. Breazeale and translated by R.J. Holingdale. Cambridge and New York: Cambridge University Press.

Reddy, W. 2001. *The Navigation of Feeling: A Framework for the History of Emotions.* New York: Cambridge University Press.

Reynolds, H. 2010. Are nations made in war? In M. Lake, H. Reynolds with M. McKenna and J. Damousi (eds) *What's Wrong with Anzac? The Militarization of Australian History*. Sydney: New South/UNSW Press, 24–44.

Rosenwein, B.H. 2002. Worrying about emotions in history. *American Historical Review* 107 (3): 821–845.

Ward, R. 1958. *The Australian Legend*. Melbourne: Oxford University Press.

Witcomb, A. 2013. Understanding the role of affect in producing a critical pedagogy for history museums. *Museum Management and Curatorship* 28 (3): 255–271.

Witcomb, A. 2015a. Cultural pedagogies in the museum: walking, listening and feeling. In M. Watkins, G. Noble and C. Driscoll (eds) *Cultural Pedagogies and Human Conduct*. London: Routledge, 158–170.

Witcomb, A. 2015b. Toward a pedagogy of feeling: understanding how museums create a space for cross-cultural encounters. In A. Witcomb and K. Message (eds) *Museum Theory*. Part of *The International Handbooks of Museum Studies*, S. MacDonald and Rees Leahy, R. (general eds). Malden/Oxford: Wiley Blackwell, 321–344.

Zeldin, T. 1971. Social history and total history. *Journal of Social History*, tenth anniversary issue: Social History Today and Tomorrow? 10 (2): 237–245.

Zeldin, T. 1982. Personal history and the history of the emotions. *Journal of Social History*, special issue on the History of Love, 15 (3): 339–347.

13 Witnessing and affect

Altering, imagining and making spaces to remember the Great War in modern Britain

Ross Wilson

This chapter uses non-representational theory to examine the way in which the popular memory of the Great War (1914–1918) in contemporary Britain is an emotional engagement formed through a process of 'witnessing' the past. This mode of public connection with the conflict has been criticized by some scholars for its reliance on representations in the media. However, such assessments obscure the emotive bond with the war's remembrance that has been formed from the cessation of hostilities to the present day. In the aftermath of the Great War, vast cemeteries and imposing monuments were built across the former battlefields; this commemorative landscape was replicated across Britain, as cities, towns and villages saw the erection of local memorials. National sites of memory at the Cenotaph in Whitehall and the Tomb of the Unknown Warrior in Westminster Abbey were also constructed as a response to the scale of death. These sites of remembrance became significant for the bereaved as places of pilgrimage while national commemorative activities imbued these locales with meaning for wider society. A century after the outbreak of the war, these spaces of memory remain central in the recollection of the war dead. However, with the passing of the last of the veterans, the conflict is now removed from 'living memory' and alternative spaces of remembrance have been formed for contemporary society.

Since the 1990s, a new wave of memorials to the 'Pals Battalions', 'Football Battalions' or specific individuals have accompanied an expansion of the museological display of the war. These sites of memory have been constructed by a society that is chronologically distant from the direct experience of the war as a means of maintaining the emotional significance of the conflict in the present day. As such, these arenas are instructional spaces; they are used to create 'witnesses' through informing current generations about the importance of a war fought a hundred years previously. To act as a witness places moral, social and political obligations onto the individual through an emotional connection, as they are required to bear the burden of memory and to testify to its significance. Therefore, this chapter will examine the way in which contemporary British society is asked to serve as 'witnesses' to the conflict through a detailed assessment of the structure and content of these new spaces of remembrance. In this manner, the emotional connection to the war is regarded not as a barrier to understanding, nor as a simple consumption of representations, but as a dynamic process that creates meaning.

Emotion and affect: witnessing the Great War

The war of 1914–1918 and its remembrance has become a contested issue for historians, politicians and the media in Britain (see Bond 2002). These debates have focused on the contemporary meanings and purpose of a conflict fought at the outset of the twentieth century. Whereas scholars have highlighted the persistence of a 'social memory' of the war that focuses on pity, suffering and loss, revisionist historians have asserted notions of advancement, development and victory as key characteristics for Britain in this first global conflagration (Todman 2005). This disparity is often assessed to be the product of the conflict's emotive representation within popular culture, in memoirs, novels, film, television and drama from the 1920s to the present day (after Hanna 2009). These accounts have been critiqued for their clichéd assessments of the war that do not advance beyond the affecting image of official incompetency or the atrocious conditions of the battlefields that are aptly reflected in the perceived 'mud, blood, rats and gas' of the Western Front (see Corrigan 2003). In contrast, historians have demonstrated that the conflict can be considered as a success, with high levels of morale at home and at the front, an unprecedented level of mobilization within industry, the economy as well as wider society and tactical advances throughout the four years of war (Sheffield 2002; Williams 2009). In this assessment, this 'forgotten victory' has been obscured by the writings of disillusioned officers in the 1920s such as Siegfried Sassoon (1886–1967) or Robert Graves (1895–1985), politically motivated representations on television and film from the 1960s such as *Oh! What a Lovely War* (1969) and, later, *Blackadder Goes Forth* (1989) and formulaic novels and dramas during the 1990s such as *Regeneration* (Barker 1991) or *Birdsong* (Faulks 1993). By analysing these media, the 'popular memory' of the conflict in Britain is frequently assessed as deficient for its reliance upon emotional imagery and reactions (see Badsey 2001).

However, this obscures the way in which emotional responses have structured the remembrance of the war and still persist within British society. Indeed, to speak of 'the trenches', 'no man's land', 'going over the top', Gallipoli, the Somme, Passchendaele or Ypres serves to automatically conjure a strong affective response based on notions of the pity, suffering and trauma of the war (Wilson 2013). To analyse these attitudes and the popular memory of the war as based solely on media representations prevents the assessment of how practices of memory are organized, performed and ordered within society (see Wertsch 2002). Therefore, non-representational theories that move beyond a focus upon the media provide an alternative means of engagement with the 'popular memory' of the war (see Thrift 2008). Rather than assume that representations are vapidly consumed by individuals, groups and communities, the remembrance of the war can be more accurately regarded as organized and performed for affect and effect (see Lorimer 2005). In this approach, the role of the witness is paramount within non-representational theory (Dewsbury 2003). While the function of the 'witness' is firmly established within a legal context, the place of the witness within Judeo-Christian culture reveals a focus on active participation (Ricoeur 2004: 264–265). The 'witness' in this context is not an objective recorder but a figure whose perception is acknowledged to be specific in time, individual in scope and singular in experience (after Lyotard 1988: 26–27). What is significant about the witness is their role towards the event; the manner in which their attitudes, ideas, values and identities are formed in relation to the action with which they have engaged (Thrift 2000). To study the 'witness perspective' requires an analysis of action and agency not passive consumption (Thrift 2003).

If the popular memory of the war is regarded as a 'series of becomings' (after Deleuze and Guattari 1988), where the remembrance of the past is performed and reinterpreted in response to representations rather than derived from them, then the role of the witness in commemoration can be explored. By examining the development of memory within Britain from the Armistice to the present day, the effect of such actions can be assessed as a means of forming identities (after Black 2004). National, regional, political, familial and moral notions of self and community are constructed through the performances of memory in relation to the conflict. Undeniably, these acts have been altered and conditioned by the operation of power within society but the nature of the role of the witness as a specific engagement ensures that it can be not only the function of authority but also the centre of resistance (after Smith 2006). In essence, the place of the witness revolves upon the notion of testimony. Witnessing can call the individual to testify to their experiences and to recognize their effect while witnessing can also reduce the individual to an observer functioning solely as the bearer of knowledge (Dewsbury 2003). The latter can be observed to constitute a 'passive witnessing' and the former an 'active witnessing'. Both roles demand that the individual should bear the burden of memory. The difference between the two positions is that the latter requires the witness to testify as to the effect of this responsibility in the present, to ensure that the act of witnessing has purpose beyond notions of knowledge and awareness (Dewsbury 2003). The creation of this witness perspective in Britain, from the initial post-war period to the commemoration of the centenary of the outbreak of war in 2014, demonstrates the way in which emotion and affect have transformed spaces of commemoration.

Post-war performances of memory

The 'battlefield of memory' that has formed with regard to the First World War has been fought over since the signing of the Armistice in November 1918. The denouement of the conflict in November 1918 was marked by debates regarding how such a war, that wrought the deaths of over 700,000 individuals from Britain, could be remembered. With such a vast scale of mourning, the commemoration of the conflict was organized as a state-led initiative. While this has been observed to be the only recourse by the national government to the recruitment and conscription of a 'civilian army', it was also undertaken as a means to ensure a degree of control over the witnessing of the conflict for wider society (see Cannadine 1981). Therefore, an official committee of civil servants and appointed architects formed under the organization of the Imperial War Graves Commission (IWGC) envisioned the war as a sacrifice for God, King and Empire and relayed this perception through a grand memorial landscape of monuments and cemeteries across the former battlefields (Heffernan 1995). In Britain, the harnessing of private grief for public service was evidenced in the construction of the Cenotaph located at the heart of government in Whitehall and the interment of the 'Unknown Warrior' in Westminster Abbey (Gregory 1994). Alongside these national memorials, local memorials to the service and sacrifice of the dead were erected by veterans, businesses and subscriptions to local societies on village greens, town squares or besides the busy city thoroughfares to act as the focal point dedications to the nation (see Gaffney 1998). With the development of a two-minute silence from the first anniversary in 1919 and the use of poppies as a marker for remembrance from 1921, the commemoration of the war dead in Britain was effectively nationalized as a public witnessing where testimonies were noticeably absent (see King 1998).

Despite the state organization of this witnessing, the local and national points of remembrance proved to be popular. Through the actions of individuals in response to these structures, the act of witnessing forged a sense of place. Sites such as the Menin Gate, located in the Flemish town of Ieper (Ypres), unveiled in 1927 as a dedication to the over 55,000 'missing' soldiers who died in the area with no known grave, became a significant site of remembrance through these performances of witnessing. The memorial book at the site (CWGC Reports 1936) records a consistently high number of individuals from Britain attending the monument:

August 1928: 14,864
August 1929: 15,174
August 1930: 13,814
August 1931: 13,416
August 1932: 9,063
August 1933: 11,390
August 1934: 11,150
August 1935: 12,015
August 1936: 18,832

In this 'battlefield pilgrimage' the conception of the memorial landscape as a particular sense of place can be observed (Lloyd 1998). The cemeteries, memorials and monuments of the former battlefields, envisioned initially as the evocation of national and imperial endeavour, began to be observed as the place of personal and familial connection. Visitors to these locations would trace the name of a loved one upon the memorial or regard the headstone as their particular site of memory and mourning. Sir Fabian Ware (1869–1949), founder of the IWGC, reflected upon this process just as the memorial landscape had been completed: '[A]round them [the memorials] is steadily growing up the feeling of intimate public and individual ownership' (Ware 1928: 3). The same process of imbuing memorials with 'public sentiment' can be observed at a local level within Britain as attachments to these places of commemoration by the bereaved and grief-stricken are formed through usage and engagement (after Connelly 2002; Stephens 2007). In the example of the northern English town of Clitheroe, donations from the public provided the capital necessary to acquire the grounds surrounding the medieval castle upon which a sculpture of a British soldier was commissioned from the artist Louis Frederick Roslyn (1878–1934). The piece, cast in bronze, is of a solemn soldier, with his head bowed as a mark of respect for the dead and was finally unveiled in 1923 and became the centre of the town's commemorative activities and for local people to grieve (see Moriarty 1995).

The spaces of commemoration for the conflict developed in the aftermath of the war are places of witnessing that encompass not only official objectives but also individual concerns. This is significant as frequently the memorials and monuments to the war are interpreted as fixed points, which control ideas about the conflict and the soldiers and civilians who experienced its effects (Bushaway 1992). Contrary to this approach, these sites of memory function as a means through which individuals, groups and communities place themselves in relation to the past and the present. This demonstrates the continuing role of these war memorials and monuments in society to the current day (see Iles 2008). Tours to the battlefields in France, Flanders and Turkey remain highly popular within contemporary society while campaigns to preserve and protect local war memorials attract considerable support (Iles 2003). Through this process, the memorials and commemorative practices have been maintained within Britain as not merely a demonstration of nationalism but also as a means to bear witness to personal, community, national and historical trauma.

Modern memory of the First World War in Britain

The post-war memorial landscape across the battlefields and the commemorative sites in Britain provided a space for witnesses to the war. However, with the passing of time and the deaths of the veterans of the conflict, this act of witnessing has been performed within new sites as a means of testifying for alternative purposes. This was notably observed with the opening of the Island of Ireland Peace Park in Mesen (Messines), near Ieper (Ypres) in 1998 (Graham and Whelan 2007). This commemorative space served to ensure the witnessing of the conflict as a shared cultural trauma as a means of reinforcing the contemporary reconciliation process (Iles 2006). The creation of this new site of mourning reflects the alterations that have occurred since the 1990s in Britain as other means and methods of remembrance have been formed through commemorative sites. This process has occurred at a particular juncture within the remembrance of the war as the last of the veterans have passed away resulting in a sense of the war slipping from 'living memory' (see Dyer 1994). However, what can be discerned in this process is the manner in which new acts of witnessing have altered the perception of the war. What is distinctive about this role of the witness is the way in which this commemorative act is used to build connections to the present.

The recent development of new spaces of memory to commemorate the First World War can be most clearly observed with the development of memorials to the 'Pals Battalions' (see Furlong et al. 2002). These military units were raised in the 'rush to the colours' during the outbreak of war in August 1914; their particular mobilization in which they enabled men of the same town or profession to serve together enabled the cultivation of morale and camaraderie. However, this also resulted in catastrophic consequences for communities if the battalion suffered heavy losses in engagements such as Gallipoli (1915) and the Battle of the Somme (1916) (see Moorhouse 1992). In the 1980s, the rise of local and family history encouraged the interest in these battalions while popular representations of the 'pals' in the wider media emphasized the emotional connection to the past (after Winter 2000). These accounts frequently highlighted the trauma of the deaths of friends while serving in the trenches and the grief of their families on the home front (see Whelan 1982). This connection to the 'Pals Battalions' led to the development of new memorials in Britain and on the former battlefields in France and Flanders to cement the association between past and present. For example, one of the first of this new commemoration was the memorial to the Liverpool and Manchester Pals at Montauban on the Somme, erected in 1994. Taking the form of a simple memorial stone, the cap badges of the Liverpool and Manchester regiments are carved alongside the inscription: 'To the glorious memory of the Liverpool and Manchester Pals who as part of the 30th Division liberated this village 1 July 1916.' The campaign for the memorial was developed during the early 1990s, after the work of determined local historians to promote the history of those who had served in the war from the Merseyside area (Maddocks 1991). Funds were raised from the area, designers were commissioned and the land was purchased without assistance of the official bodies responsible for the maintenance of the memory of the war. The work was undertaken in the belief that the 'pals' from the north-west region of England were a 'special breed' and that the sacrifices of those from the two great industrial cities had not been sufficiently regarded (Maddocks 1999: 150). Visitors to the site were, therefore, called upon to witness the war as an expression of regional identity. Such uses of witnessing the past to serve the interests of present identities can also be observed in the construction of other monuments to the 'Pals Battalions' that have been erected within the past two decades. For example, after the death of the last remaining veteran of the 'Barnsley Pals' who passed away in the early 1990s and the revival of local interest in the unit's history, a campaign was launched to build a memorial through public donations. The village of Serre on the Somme was chosen and a black granite tablet was unveiled in 1998 within the Sheffield Memorial Park, a site that was initially inaugurated as a commemorative space in the 1930s.

The Sheffield Memorial Park is evidence itself of this new wave of commemoration as it has become the site of memorials constructed within the past 20 years (after Gough 1998, 2001). 'Pals Battalion' memorials were erected here during the 1990s as regions within Britain sought to establish their connection to the conflict. One of the most prominent examples of this process was the commemoration of the 'Accrington Pals' (Jackson 2013). This unit was one of the first to be the subject of a dedicated local history assessment that encouraged the construction of a memorial, funded by local townspeople from Accrington in north-west England, which was placed in Sheffield Memorial Park in 1991 (see Turner 1987, 1998). The memorial, built in the distinctive red brick from the Accrington area, is dedicated to all members of that battalion who were killed during the Battle of the Somme. Focused on the tragedy of the Battle of the Somme, which saw devastating losses and casualties on the first day of operations, these memorial sites evoke an emotional response to the trauma of the war (Dunkley et al. 2011). This, however, constitutes an active witnessing as visitors to the areas are required to testify to the loss and significance of these deaths from their own locale. Such processes can be observed with the 2006 unveiling at Sheffield Memorial Park of a memorial plaque to the Burnley Pals, which was specifically stated as an act of witnessing: 'We are all very proud to see the memorial unveiled in tribute to these brave men. Around 300 to 400 Burnley soldiers were killed on the Somme, and this plaque is a fitting tribute to their sacrifice' (Anon. 2006).

Such affective points of association to the past enable the representation of both regional and national identity within Britain. This can also be observed with the unveiling in 2007 of a Celtic cross monument to Scottish soldiers who died during the Battle of Passchendaele (1917), which was erected at the Frezenberg Ridge with funding from the Scottish Government. The cross was intended to mark the distinctive contribution of Scotland's soldiers at the front. This can also be noted in the sculpture of a Welsh Dragon tearing at barbed wire, which was erected in 1987 near Mametz Wood in the Somme Valley in memory of the 38th (Welsh) Division. This unit, which was comprised largely of Welshmen from the Royal Welch Fusiliers and the South Wales Borders, suffered large-scale fatalities during the Somme offensive and the sculpture is intended to ensure the acknowledgement of the specific deaths of soldiers from Wales (Gough 1998).

Such emotional and affective engagements with the past are also evidenced in the new memorials to the 'Pals Battalions' that have been unveiled in Britain from the 1990s as new associations to the First World War are constructed. Recent tablets, friezes and sculptures in towns and cities such as Manchester, Bradford, Preston and Chorley have sought to draw attention to the emotional trauma of the war (see Wilson 2013: 8–10). These memorials are usually placed in distinct locations within the urban landscape: town centres where volunteers gathered in August 1914; streets where the soldiers paraded before being sent for training; or railway stations where fond, and possibly final, farewells were exchanged between loved ones (after Tarlow 1997). The emotional intensification of these sites provides further evidence of how such sites are used by communities. For example, the Liverpool Pals Memorial was unveiled at Liverpool Lime Street Railway Station in August 2014; these bronze friezes, depicting the separation and loss wrought by the war on the city's populace, were located where soldiers would have departed for the front lines. The chairman of the Liverpool Pals Memorial Fund, organized after the new millennium to raise donations for the commemorative artwork, Lt Col (Retd) Anthony Hollingsworth reiterated the evocative place of the friezes and their function for civic and wider society:

> One hundred years ago to this day, these would-be Pals lined the streets of Liverpool and St George's Hall to enlist. By 10am on that first morning the first battalion had formed. By September 7, three full battalions had formed. We want to ensure the Pals' story is remembered and becomes part of the fabric of this city and beyond.
>
> (Cited in Jones 2014)

The location of these memorials to the 'Pals Battalions' at sites of trauma and tragedy not only emphasizes the sense of loss associated with the war but also reinforces a sense of place for contemporary communities (after Dawson 2005). This is particularly observable in the commemoration of the 'Pals Battalions' associated with football teams from Britain and with the advent of the centenary of the conflict these specific connections have been strengthened (see Wilson 2014). Memorials to 'Football Battalions' have been constructed in France and Belgium over the past decade. For example, the Contalmaison Cairn on the Somme was unveiled in 2004 to commemorate the losses suffered by the 'Sporting Battalion', a unit raised in Edinburgh and largely comprised of footballers from Heart of Midlothian Football Club. This memorial, paid for with funds raised by the public, was designed to ensure that Scottish materials and Scottish craftsmen could mark the deaths of Scottish soldiers who died during the assault conducted as part of the first day of the Battle of the Somme (McCrae's Battalion Trust 2014). In 2010, the Football Battalion (17th Middlesex Regiment), which was recruited in 1914 from the ranks of professional football clubs, was commemorated near the village of Longueval on the Somme where it also suffered significant losses during the battle. In the last few years, West Ham Football Club (in 2009) and Portsmouth Football Club (in 2014) have both unveiled memorial plaques at their grounds to remember the deaths of their players during the First World War. These poignant reminders of young lives cut short before establishing their careers as players serves to reinforce the sense of loss and tragedy associated with the conflict of 1914–1918. Such associations were also present in the construction of a memorial to Walter Tull (1888–1918), the professional footballer who became an officer in the British Army during the First World War (Vasili 1996). This structure, placed outside Northampton Town's Sixfields Stadium, Tull's last club before being posted to the battlefields, is inscribed with a call for visitors to witness how a man 'rendered breathless in his prime' can be regarded as a symbol of a struggle for equality. National, regional, cultural and political identities are thereby constructed through the performances of witnessing at these sites.

Such acts of witnessing are also present in the recent museum exhibitions launched to mark the centenary of the outbreak of the war (Whitmarsh 2001). From February 2014, the Museum of Lancashire (2014), based in Preston, offered an opportunity to experience the 'sights and smells' of the front lines in their display 'Lancashire at War'. In this immersive exhibition, visitors were guided through recreated trenches, replete with artefacts of the war, to enable individuals to grasp the physical experiences of the battlefields (see Winter 2012). This manner of museological engagement has been an increasing feature of First World War exhibitions since the development of the Imperial War Museum's (IWM) 'Trench Experience' in 1989 (Espley 2008). This particular display enabled the visitor to walk through the trenches before a raid was launched on the enemy's lines; the heavy artillery fire, darkness and confined space was included to facilitate the understanding of the past for present-day visitors (Borg 1991).

Following this development, institutions such as the Museum of the Manchester Regiment (2014) in Ashton and the Royal Engineers Museum (2014) in Gillingham have constructed their own 'trenches' within displays as an educational device. This approach has been employed by a number of regional museums in response to the advent of the centenary. In the South Wales town of Porthcawl, the local museum opened its own 'Trench Experience' in April 2014, which utilized the timbers of a house from the last century to recreate a 'life-size' trench exhibition with artefacts donated by local townspeople linking the trenches to the homefront and the past to the present (Porthcawl Museum 2014). The Staffordshire Regimental Museum (2014) in Whittington renovated its outdoor trench display with the approach of the centenary. The initial display was constructed in 2002, but adjustments were made in 2014 to ensure that the exhibition represented a more 'realistic' demonstration of life during the First World War. Further trench recreations in local and regimental museums such as the York Castle Museum (2014), the Museum of the Royal Leicestershire Regiment (2014), the Wycombe Museum (2014) and the Saltash Museum and Local History Centre (2014) highlight the significance of witnessing this specific experience of the war for contemporary audiences. The performances of visitors to these 'trench experiences' constitutes an active witnessing, as, while dislocated from the reality of industrialized warfare, the physical and emotional engagement with the environment ensures a sense of place is formed for visitors. This witnessing is inevitably particular and restricted, but it is through these performances that exhibitions acquire meanings and values for their audiences (see Smith 2007). Indeed, such was the close association between visitors and the 'trench experience' at the IWM that fears were expressed within the media after plans for the museum's 2014 refurbishment were unveiled that showed this 'resource' would be lost (see Kennedy 2014).

The recent development of museums representing the history and experience of the First World War have been part of wider commemorative practices that have developed to mark the centenary of the outbreak of the conflict in Britain. These national and local initiatives have seen the creation of permanent and temporary spaces of remembrance to engage and inform current society with the conflict of 1914–1918. One of the more prominent memorial acts was undertaken by the Guards Museum (2014) in London to build a memorial garden at the institution with the support of Flanders House in London, the Commonwealth War Graves

Commission and the Belgian-Luxembourg Chamber of Commerce in Great Britain (Memorial Garden 2014). Entitled 'Flanders Fields, 1914–2014', the garden has been formed through the use of soil from the battlefields in Flanders, which was collected in sandbags by British and Belgian schoolchildren and solemnly paraded through London (Anon. 2014). The procession of earth and its part in the landscaping of the garden was intended as a highly affective and symbolic act to honour the lives and deaths of British soldiers and the ongoing relationship between Britain and Belgium. The emotional nature of this new space of memory was noted by the curator of the Guards Museum Andrew Wallis as he described how this seemingly 'sacred earth' would remain 'in sight of the main drill square from where so many guardsmen marched away never to return' (cited in Anon. 2014). Officially opened in November 2014, the memorial garden's unveiling by Queen Elizabeth II was attended by dignitaries and served to instil a sense of witnessing the past to inform the present. Indeed, Major General Edward Smyth-Osbourne, the General Officer Commanding the Household Division stated:

> The Guards fought in almost every battle of the First World War. This memorial garden stands proud testament to their achievements in what Winston Churchill called 'the world crisis', and is testament to the traditions that we all strive to live up to today.
>
> (HM Govt 2014a)

This value of witnessing as a civic duty is embedded within the recent commemorative activities associated with the marking of the centenary of the war. New spaces of commemoration have been constructed as instructional locales where the values and ideals of the past can inform and invigorate contemporary society. This is evidenced in the Government-led programme announced in August 2013 to construct paving stones in the villages, towns and cities of those awarded the Victoria Cross during the First World War. The highest military decoration for valour in the face of the enemy was awarded to over 600 individuals during the conflict, including individuals from Britain and those from other nations who fought for the British Army. The objectives of the campaign within Britain appear to rely upon affect to ensure engagement from local communities (HM Govt 2014b). The stated aim of the project was to:

- honour their bravery;
- provide a lasting legacy of local heroes within communities; and
- enable residents to gain a greater understanding of how their area fitted into the First World War story.

The public virtue of witnessing the commemoration of the Victoria Cross winners is emphasized in this campaign as contemporary society is emotionally engaged with the service and sacrifices of their forebears. A similar process was enacted with the candlelight vigil within churches across Britain and the promotion of 'Lights Out', a movement to switch all lights off in homes except for a single candle, to mark the centenary of the declaration of war on 4 August 2014. A temporary space of commemoration was created in this activity, which brought individuals and communities together within a collective act of remembrance. Inspired by the phrase attributed to the British Foreign Secretary Lord Grey that the 'lamps were going out all over Europe', this memorial practice also forged a sense of witnessing that was divorced from testimony. Marking and recognizing the passing of an event offered no means of reflection or sense of place only the notion that such witnessing was conducted as a 'service' or as 'recognition'.

This passive witnessing was most evidently displayed in the installation placed around the moat surrounding the Tower of London from July to November 2014. This piece, entitled *Blood Swept Lands and Seas of Red*, saw the planting of over 800,000 ceramic poppies to mark the deaths of British and Commonwealth soldiers during the war. The spectacle attracted vast numbers of visitors and drew attention across the media for the way in which it evoked a sense of gravity and significance amongst observers. However, the space of the artwork placed the individual into the role of a passive witness; called upon to recognize the scale of death and drawn into this emotional engagement, the witness is required to make no testimony and to claim no association. The temporary nature of the installation further demonstrated how the official marking of the centenary of the war in Britain has rendered the position of the witness into that of the neutral observer. However, the response to this installation from visitors demonstrates how the role of the witness can subvert the organization of memory. The outcry over the dismantling of the artwork, campaigns to ensure its survival and the way in which a sense of place was formed in response to the piece, transformed the field of ceramic poppies just as the commemorative landscape of the Western Front had been altered from its initial conception as a statement of national power to an intimate portrayal of loss and bereavement. To serve as a witness to the conflict places the individual in a particular relationship to the past; it is the performance of this act of witnessing that transforms representations and creates the 'popular memory' of the war rather than the representations themselves that transform public memory.

Conclusions

What has marked the remembrance of the war since the Armistice has been the individual acts of witnessing that have shaped the relationship between the historical event and the society that honours it. Through an engagement with non-representation theories, which privilege the performances, actions and responses of individuals, groups and communities, rather than the representations that are presumed to structure actions, the manner in which the remembrance of the First World War has been enacted across Britain can be examined. The passing of the last veteran and the commemoration of the centenary of the advent of the war has not dimmed the emotional effect of the conflict. Contemporary society still performs an emotive act of witnessing the past as a process of forming a sense of place and identity. Whether regional, national, moral or political, the witness can associate and define themselves through a connection with the conflict through an act of testifying. The recent memorials that commemorate particular 'Pals Battalions' enable an expression of regional identity, while museum displays provide an emotional engagement with the physical conditions of the war. While the acts of witnessing can be both active and passive, requiring visitors to recognize the effect of the war through an act of testimony or serving as a neutral observer and bearer of memory, these acts still constitute a sense of becoming. Rather than dismissing the emotional nature of this commemoration as inaccurate 'myths', it can be regarded as a vital means by which a relationship to the past is understood.

References

Anon. 2006. Somme sight for plaque to honour fallen. *Burnley Express*, 10 July. [Online]. Available from: www.burnleyexpress.net/news/local/somme-sight-for-plaque-to-honour-fallen-1-1670375, last accessed 3 December 2014.

Anon. 2014. Sacred soil from Flanders Fields & the Memorial Garden. *The Guards Magazine*. [Online]. Available from: http://guardsmagazine.com/features_flandersfields. html, last accessed 4 December 2014.

Badsey, S. 2001. *Blackadder Goes Forth* and the 'two Western Fronts' debate. In G. Roberts and P.M. Taylor (eds) *The Historian, Television and Television History*. Luton: University of Luton Press, 113–125.

Barker, P. 1991. *Regeneration*. London: Penguin.

Black, J. 2004. Thanks for the memory: war memorials, spectatorship and the trajectories of Commemoration 1919–2001. In N. Saunders (ed.) *Matters of Conflict: Material Culture, Memory and the First World War*. London: Routledge, 134–148.

Bond, B. 2002. *The Unquiet Western Front: Britain's Role in Literature and History*. Cambridge: Cambridge University Press.

Borg, A. 1991. New developments at the Imperial War Museum. *Interpretation Journal* 47: 6–7.

Bushaway, B. 1992. Name upon name: the Great War and remembrance. In R. Porter (ed.) *Myths of the English*. Cambridge: Polity Press, 136–167.

Cannadine, D. 1981. War and death, grief and mourning in modern Britain. In J. Whaley (ed.) *Mirrors of Mortality: Studies in the Social History of Death*. New York: St Martin's Press, 187–242.

Connelly, M. 2002. *The Great War, Memory and Ritual: Commemoration in the City and East London, 1916–1939*. Woodbridge: Royal Historical Society and Boydell Press.

Corrigan, G. 2003. *Mud, Blood and Poppycock*. London: Cassell.

CWGC Reports. 1936. *Commonwealth War Graves Commission Reports*, 219/2/1.

Dawson, G. 2005. Trauma, place and the politics of memory: Bloody Sunday, Derry, 1972–2004. *History Workshop Journal* 59 (1): 151–178.

Deleuze, G. and Guattari, F. 1988. *A Thousand Plateaus: Capitalism and Schizophrenia*. Trans. by B. Massumi. London: Continuum.

Dewsbury J.-D. 2003. Witnessing space: 'knowledge without contemplation'. *Environment and Planning A* 35 (11): 1907–1932.

Dunkley, R., Morgan, N. and Westwood, S. 2011. Visiting the trenches: exploring meanings and motivations in battlefield tourism. *Tourism Management* 32 (4): 860–868.

Dyer, G. 1994. *The Missing of the Somme*. London: Phoenix Press.

Espley, R. 2008. 'How much of an "experience" do we want the public to receive?': trench reconstructions and popular images of the Great War. In J. Meyer (ed.) *British Popular Culture and the First World War*. Leiden: Brill, 325–350.

Faulks, S. 1993. *Birdsong*. London: Vintage.

Furlong, J., Knight, L. and Slocombe, S. 2002. They shall grow not old: an analysis of trends in memorialisation based on information held by the UK national inventory of war memorials. *Cultural Trends* 12 (45): 1–42.

Gaffney, A. 1998. *Aftermath: Remembering the Great War in Wales*. Cardiff: University of Wales Press.

Gough, P. 1998. War memorial gardens as dramaturgical space. *International Journal of Heritage Studies* 3 (4): 199–214.

Gough, P. 2001. Landscapes of war (and peace). In R. White (ed.) *Monuments and the Millennium*. London: James & James and English Heritage, 228–236.

Graham, B. and Whelan, Y. 2007. The legacies of the dead: commemorating the Troubles in Northern Ireland. *Environment and Planning D* 25 (3): 476–495.

Gregory, A., 1994. *The Silence of Memory: Armistice Day 1919–1946*. Oxford: Berg.

Guards Museum. 2014. Homepage. [Online]. Available from: www.theguardsmuseum.com/, last accessed 6 December 2014.

Hanna, E. 2009. *The Great War on the Small Screen: Representing the First World War in Contemporary Britain*. Edinburgh: Edinburgh University Press.

Heffernan, M. 1995. For ever England: the Western Front and politics of remembrance in Britain. *Ecumene* 2 (3): 293–323.

HM Govt. 2014a. Flanders' Fields Memorial Garden opens. [Online]. Available from: www.gov.uk/government/news/flanders-fields-memorial-garden-opens, last accessed 12 December 2014.

HM Govt. 2014b. Bringing people together in strong, united communities. Homepage. [Online]. Available from: www.gov.uk/government/policies/bringing-people-together-in-strong-united-communities/supporting-pages/victoria-cross-commemorative-paving-stones, last accessed 4 November 2014.

Iles, J. 2003. Death, leisure and landscape: British tourism to the Western Front. In M. Dorrian and G. Rose (eds) *Deterritorialisations: Revisioning Landscapes and Politics*. London: Black Dog, 234–243.

Iles, J. 2006. Recalling the ghosts of war: performing tourism on the battlefields of the Western Front. *Text and Performance Quarterly* 26 (2): 162–180.

Iles, J. 2008. Encounters in the fields: tourism to the battlefields of the Western Front. *Journal of Tourism and Cultural Change* 6 (2): 138–154.

Jackson, A. 2013. *Accrington's Pals: The Full Story*. Barnsley: Pen and Sword.

Jones, C. 2014. Earl of Wessex unveils the Liverpool Pals memorial at Lime Street Station. *Liverpool Echo*, 31 August.

Kennedy, M. 2014. Imperial War Museum's new look at a century of warfare. *The Guardian*, 16 July.

King, A. 1998. *Memorials of the Great War in Britain: The Symbolism and Politics of Remembrance*. Oxford: Berg.

Lloyd, D.W. 1998. *Battlefield Tourism: Pilgrimage and Commemoration of the Great War in Britain, Australia and Canada 1919–1939*. Oxford: Berg.

Lorimer, H. 2005. Cultural geography: the busyness of being 'more-than representational'. *Progress in Human Geography* 29: 83–94.

Lyotard, F. 1988. *Le Différend*. Minneapolis: University of Minnesota Press.

McCrae's Battalion Trust, 2014. Homepage. [Online]. Available from: www.mccraesbattalion trust.org.uk, last accessed 12 December 2014.

Maddocks, G. 1991. *The Liverpool Pals: A History of the 17th, 18th, 19th, and 20th (Service) Battalions The King's Liverpool Regiment, 1914–1919*. London: Leo Cooper.

Maddocks, G. 1999. *Montauban*. Barnsley: Pen and Sword.

Memorial Garden. 2014. Homepage. [Online]. Available from: www.memorial2014.com/ en/memorial_garden, last accessed 4 December 2014.

Moorhouse, G. 1992. *Hell's Foundations. A Town, Its Myths and Gallipoli*. London: Hodder & Stoughton.

Moriarty, C. 1995. The absent dead and figurative First World War memorials. *Transactions of the Ancient Monuments Society* 39: 8–40.

Museum of Lancashire. 2014. Lancashire at War. [Online]. Available from: www. lancashire.gov.uk/leisure-and-culture/museums/museum-of-lancashire.aspx, last accessed 4 December 2014.

Museum of the Manchester Regiment. 2014. Homepage. [Online]. Available from: www. tameside.gov.uk/museumsgalleries/mom, last accessed 4 December 2014.

Museum of the Royal Leicestershire Regiment. 2014. Newarke Houses Museum and Gardens. [Online]. Available from: www.leicester.gov.uk/your-council-services/lc/ leicester-city-museums/museums/newarkehouses/, last accessed 4 December 2014.

Porthcawl Museum. 2014. Homepage. [Online]. Available from: www.porthcawlmuseum. com/, last accessed 4 December 2014.

Ricoeur, P. 2004. *Memory, History, Forgetting*. Chicago: University of Chicago Press.

Royal Engineers Museum. 2014. Homepage. [Online]. Available from: www.re-museum. co.uk/, last accessed 4 December 2014.

Saltash Museum and Local History Centre. 2014. Homepage. [Online]. Available from: www.saltash-heritage.org.uk/, last accessed 4 November 2014.

Sheffield, G. 2002. *Forgotten Victory: The First World War: Myth and Reality*. London: Headline.

Smith, J.G. 2007. Learning from popular culture: interpretation, visitors and critique. *International Journal of Heritage Studies* 5 (3–4): 135–148.

Smith, L. 2006. *Uses of Heritage*. London: Routledge.

Staffordshire Regimental Museum. 2014. Homepage. [Online]. Available from: http:// staffordshireregimentmuseum.com/, last accessed 4 December 2014.

Stephens, J. 2007. Memory, commemoration and the meaning of a suburban war memorial. *Journal of Material Culture* 12 (3): 241–261.

Tarlow, S. 1997. An archaeology of remembering: death, bereavement and the First World War. *Cambridge Archaeological Journal* 7: 105–121.

Thrift, N. 2000. Afterwords. *Environment and Planning D: Society and Space* 18: 213–255.

Thrift, N. 2003. Performance and. . . . *Environment and Planning A* 35: 2019–2024.

Thrift, N. 2008. *Non-Representational Theory: Space, Politics, Affect*. Abingdon and New York: Routledge.

Todman, D. 2005. *The Great War: Myth and Memory*. London: Hambledon.

Turner, W. 1987. *Accrington Pals*. Barnsley: Wharncliffe Publishing.

Turner, W. 1998. *Accrington Pals Trail: Home and Overseas*. Barnsley: Pen and Sword.

Vasili, P. 1996. Walter Tull, 1888–1918: soldier, footballer, black. *Race and Class* 38 (2): 51–69.

Ware, F. 1928. Introduction. *Ninth Annual Report of the IWGC*. London: HMSO, 3–5.

Wertsch, J.V. 2002. *Voices of Collective Remembering*. Cambridge: Cambridge University Press.

Whelan, P. 1982. *Accrington Pals*. London: Methuen.

Whitmarsh, A. 2001. 'We will remember them': memory and commemoration in war museums. *Journal of Conservation and Museum Studies* 7: 1–15.

Williams, D. 2009. *Media, Memory*, and the *First World War*. Montreal and Kingston: McGill-Queen's University Press.

Wilson, R. 2013. *Cultural Heritage of the Great War in Britain*. Farnham: Ashgate.

Wilson, R. 2014. It still goes on: football and the heritage of the Great War in Britain. *Journal of Heritage Tourism* 9 (3), 197–211.

Winter, J. 2000. The generation of memory: reflections on the 'memory boom' in contemporary historical studies. *GHI Bulletin* 27: 69–92.

Winter, J. 2012. Museums and the representation of war. *Museum and Society* 10 (3): 150–163.

Wycombe Museum. 2014. Homepage. [Online]. Available from: www.wycombe.gov.uk/council-services/leisure-and-culture/wycombe-museum.aspx, last accessed 6 October 2014.

York Castle Museum. 2014. When the word changed forever. [Online]. Available from: www.yorkcastlemuseum.org.uk/exhibition/1914-when-the-world-changed-forever/, last accessed 13 December 2014.

14 Places of memory and memories of places in Nazi Germany

Joshua Hagen

Efforts to commemorate the criminality of the Nazi Party commonly generate intense scrutiny and bitter controversy, but it is often forgotten that the Nazi regime went to considerable lengths to create its own commemorative places. These places of Nazi memory sought to position the movement within a broader historical trajectory of racial struggle, heroism, sacrifice and martyrdom on behalf of the national community. Nazi ideologues harnessed this racially infused social Darwinism as a conceptual framework for interpreting the past, acting in the present and imagining the future. The Nazi Party made clear the primacy of struggle and war through its hateful speech, paramilitary appearance and violent behaviour. It was logical that these same themes permeated the movement's commemorations. Party propagandists could accomplish much through oration and writing, but the impulse to anchor Nazi memories through commemorative plaques, monuments, buildings and other types of physical markers, as well as parades, rallies and other bodily performances, proved inexorable (Mosse 1990; Koshar 1998; François and Schulze 2001; Macdonald 2006). This is hardly surprising since numerous scholars have demonstrated the power of space and place to shape collective memories and simultaneously how collective memories shape experiences of space and place (Hoelscher and Alderman 2004; Foote and Azaryahu 2007; Jones 2011; Jones and Garde-Hansen 2012a). As Alderman and Inwood noted, 'while memory is ostensibly about the past, it is shaped to serve ideological interests in the present and to carry certain cultural beliefs into the future' (2013: 187). In this context, the Nazi movement's memories and myths of racial struggle, heroism, sacrifice and martyrdom called for an iterative restructuring of Germany's commemorative typologies, spatialities and topologies.

This chapter tackles this topic by examining a range of commemorative places, spaces and practices promoted by the Nazi movement. To that end, the chapter combines foundational geographical concerns with the production of space, place, networks and movement with contemporary scholarship focusing on representations (semiotics), actions (performativity) and experiences (affect) (Dovey 2009; Anderson and Harrison 2010; Meusburger et al. 2011; Horton and Kraftl 2014). These different lines of enquiry can be positioned as theoretical opposites or at least mutually exclusive, especially when caricatured in their basest forms, when in fact much is gained from research that recognizes and harnesses their complementarity (Lorimer 2005; Whatmore 2006; Simonsen 2007; Waterton and Watson 2013). Such an effort requires a critical examination of the materiality of Nazi places of memory and memories of places beginning with the processes of selection and staging that undergirded the regime's incipient commemorative geographies. These material changes provided a tangible framework for embedding these places into Nazi memory through official and everyday discourse and representational practices. The Nazi regime also attached great importance to reinforcing approved narratives through bodily practices and performances, again ranging from spectacular to banal. Taken together, the regime endeavoured to assemble these sites/sights of Nazi memory into a geography of commemoration that would foster specific affects, emotions and experiences, and ultimately specific behaviours, supporting notions of racial struggle, heroism, sacrifice and martyrdom in service to the movement.

Scholars working in different contexts have characterized these dynamics in various ways, such 'engineering affective atmospheres' (McCormack 2008: 413) or 'a series of conditioning environments' (Thrift 2008: 236), but perhaps most useful is Adey's formulation of a 'calculative architecture of affective control' (2008: 438) since it invokes the possibilities of geographies of architecture and architectural geographies to serve as a nexus for the interplay and interconnection between semiotics, performativity and affect. Indeed, a number of fruitful forays has already been made towards more spatially nuanced theorizations of architecture (Lees 2001; Rose et al. 2010; Jacobs and Merriman 2011; Lees and Baxter 2011). Kraftl and Adey, for example, have demonstrated how architecture involves representational, performative and affective practices intended to create places and spaces preconditioned to encourage certain ways of seeing, acting and feeling, or what they termed the 'affective geographies of buildings' (2008: 228). In a similar vein, Allen (2006: 445) discusses the notion of ambient power where

> there is something about the character of an urban setting – a particular atmosphere, a specific mood, a certain feeling – that affects how we experience it and which, in turn, seeks to induce certain stances which we might otherwise have chosen not to adopt.

Much of this scholarship emphasized the immediacy of individual experience, but there have also been efforts to explore shared experiences and collective identities, especially how memories of violence and war are used to delineate group identity, inclusion and exclusion (Curti 2008; Schramm 2011; Tyner et al. 2014; Sakamoto 2015).

Drawing from this rich literature, this chapter seeks to map out 'the materialities of memories of geography and/or geographies of memories' (Jones and Garde-Hansen 2012b: 11) as imagined and enacted within the context of Nazi Germany. Admittedly, this examination requires some inference, since the historical record is limited, many places have changed considerably since the Nazi period, and other projects remained unrealized. Those caveats aside, the regime managed to establish a fairly substantial number of commemorative places, and its propagandists were hardly subtle in communicating the intended message to the public and also how people should think, act and feel in response. It is therefore possible to discern several general proclivities across the regime's commemorative practices and spatialities aimed towards (re)producing a Nazi collective identity.

Struggle was a foundational theme within the Nazis' social Darwinist worldviews, necessitating acts of heroism and sacrifice, and possibly martyrdom or murder, which blurred distinctions between street fighter and frontline soldier, political struggle and racial war (Bartetzko 1985; Baird 1990; Behrenbeck 1996; Pappert 2001). This conceptual framework attempted to place the Nazi movement within a broader struggle of good versus evil stretching back into prehistory. In a very literal sense, Adolf Hitler and his followers believed themselves to be heroic figures intervening in this tectonic (perhaps Teutonic) battle to rescue the German nation from the brink of oblivion. This was evident in Hitler's autobiographical manifesto *Mein Kampf*. The book's title is commonly translated as *My Struggle* but *Kampf* can also mean fight, conflict or battle, and in many respects, the Nazi Party functioned more like a paramilitary organization than a political party. The party's emphasis on uniforms, ranks, flags and standards brandished through highly regimented marches and rallies aimed to convey the impression of a mass movement operating beyond conventional politics. That party members routinely engaged in street brawling added a performative element to that militaristic aesthetic while also fostering a broader atmosphere of public disorder and violence. Hitler and other party members even referenced the movement's rise to power as the time of fighting or struggle (*Kampfzeit*). Those who joined the party before 1933 bore the cherished title of Old Fighter (*Alter Kämpfer*). Within this Manichean framework, places associated with seminal events in party history were reconceputalized as figurative, and in a certain sense literal, battlefields. They became constitutive components of Nazi ideology, identity and belonging. Through these places, the Nazi Party gradually amassed the basic ingredients for a commemorative liturgy complete with its own pantheon of heroes, battles, martyrs, rituals and sacred sites/sights of memory.

The regime's commemorative practices were rather decentralized and encompassed a range of initiatives that vacillated between competition and cooperation, coercion and consent. The diffuse nature of Nazi commemorative practices was both cause and effect of the regime's broad understanding of places of memory and memories of places. A popular reader for adolescents explained how Germany was 'strewn with sites of memory'. Each represented 'a piece of memory and tradition, a holy place of German heroism, German culture and spiritual splendour . . . places of pilgrimage of the nation where Germans experience the wonder of his nation and summon belief in the future' (von Schumacher 1935: 95–96). Assorted actors from cabinet ministers to local functionaries had considerable freedom to appropriate these myriad places so long as they were, in Kershaw's memorable phrasing, 'working towards the Führer' (1993). What that meant in practical terms for places of memory was therefore highly contingent, but this chapter divides the Nazi regime's burgeoning archipelago of commemorative sites/sights into three overlapping categories: those associated with the history of the Nazi Party; those tied to seminal events in German history; and finally those related to the world wars.

Placing Nazi history

Hitler was the movement's self-styled messiah, so naturally places associated with him were recast as extraordinary and even sacred soon after seizing power (Baird 1990; Bauer et al. 1993; Pappert 2001; Kriegl 2003; Taylor 2007; Heusler 2008). Commemorative plaques soon marked Hitler's first apartment in Munich, a small flat above a tailor's shop, where he lived from May 1913 until departing for military service, as well as his equally non-descript residence after the war. The Munich barracks where Hitler received his basic training were renamed Adolf Hitler Barracks. Another plaque was mounted inside the Hofbräuhaus beer hall reading: 'From this place, Adolf Hitler proclaimed on 24 February 1920 the program of the National Socialist German Workers' Party.' Hitler's hospital room in Pasewalk where he convalesced during the First World War and his later prison cell in Landsberg soon became shrine-like installations, replete with swastika flags, portraits of Hitler and other Nazi regalia. These mundane places became pivotal locations in Hitler's life story and Germany's proclaimed awakening. The party eventually purchased Hitler's birth house in the small Austrian town of Braunau am Inn and staged it as an additional place of party pilgrimage. Combined, these seemingly disparate places stitched together a background narrative of Hitler's improbable rise from humble beginnings to national saviour, while leaving other portions of his biography silent, notably his years in Vienna.

An unremarkable Munich beer hall, the Sterneckerbräu brewery (Figure 14.1), served as the inaugural meeting place of the obscure German Workers' Party in January 1919. The meetings drew small crowds, but military officials nonetheless dispatched Hitler to inform on a September gathering. Hitler soon joined the party and became its leader by its relaunch as the National Socialist German Workers' Party. Hitler and his colleagues established the first party office in one of the Sterneckerbräu's side rooms in October 1919. The party remained there for just a few months, but the building nonetheless gained fame as the Nazi Party's first public meeting venue, the site of Hitler's first encounter with the movement and the first Nazi Party headquarters, providing a solid claim to the title of birthplace of the Nazi movement. Accordingly, the Sterneckerbräu was renovated as a shrine-like museum, which Hitler opened on 8 November 1933. The dining hall featured a large painting of Hitler hung amid draping swastika flags. The opposite wall bore an eagle clutching a wreath-encircled swastika flanked by a dedication reading: 'In this corner, our Reich Chancellor Adolf Hitler founded the NSDAP on 24 February 1920. Here, the first seven brave champions laid the foundation for the German freedom movement.' Visitors could see a re-staging of the first party office purportedly furnished with the original tables, chairs, wardrobe and other mementos such as the party's first typewriter and the cigar box used as the party treasury. Assorted fliers, posters and party documents were displayed on

Figure 14.1 The Sterneckerbräu beer hall.

Source: Liese (1943: 19).

the tables and walls. The museum guidebook explained that the party's 'first major battles' revolved around this room. Highlighting the interdependencies between place, memory and affect, the museum guidebook asserted that 'one has to be there oneself in order to feel a hint from the time out of which the great movement was created' (Schüßler 1935: 5, 11).

The Nazi Party used numerous venues during the *Kampfzeit*, but the Bürgerbräukeller beer hall was one of the most frequent and provided the launching pad for Hitler's coup against the Bavarian government. After hours of indecision and hesitation throughout the night of 8–9 November 1923, Hitler and around 2,000 supporters and onlookers marched from the Bürgerbräukeller towards the district military headquarters on Ludwig Street. Police officers guarding the Ludwig Bridge confronted the column but soon backed down, allowing Hitler's entourage to continue. A second group of police officers blocked the putschists' path near the Field Generals' Hall (*Feldherrnhalle*). Gunfire erupted. Hitler's putschists scattered, leaving four police officers and 16 putschists dead. Despite failure, the putsch marked a seminal event for the Nazi movement.

Soon after taking power, the Nazi press churned out veritable hagiographies of Hitler and his followers that traced the movement's seemingly inexorable rise to power (Espe 1933; Reich and Achenbach 1933). Authors faithfully illustrated their narratives with pictures of the 'authentic' places of Nazism from the

Sterneckerbräu to the Brown House while eulogizing the putsch martyrs as the epitomes of heroism and sacrifice. The Bürgerbräukeller and the Field Generals' Hall gained a mystical aura as Nazi supporters dutifully trekked to the sites. Yet glorifying the movement through literature and imagery was insufficient; the regime also sought to (re)create tangible places and spaces of Nazi memory and commemoration. Hitler ordered a large bronze memorial topped by an eagle clutching a wreathed swastika to be erected on the Field Generals' Hall. The side overlooking the street listed the names of the martyrs who died in 'true belief in the resurrection of their nation'. The other side of the plaque bore the inscription 'And yet you have triumphed!' This allusion to victory over or through death had obvious parallels with Christian belief in Jesus' death and resurrection. An SS (*Schutzstaffel*) honour guard stood on the street below the plaque, and all passersby were expected to give the Nazi salute. In this way, Germans were made to respond to the party slogan emblazoned into this purported hallowed ground or alternatively find another route. The reimagined hall became a standard element in Nazi propaganda in an effort to keep the memories of the putsch and its martyrs alive and more broadly promulgate across Germany an atmosphere of sacrifice.

Memories of sacrifice and martyrdom also permeated the party's building projects. The Barlow Palace, purchased in 1930, was the party's first substantive architectural endeavour (Dresler 1939; Maier Hartmann 1942; Bauer et al. 1993; Heusler 2008). Located near the famed King's Square (*Königsplatz*), the palace was built in 1828 in a Biedermeier style that lent the rabble-rousing Nazis a veneer of respectability. The location was also beneficial since King's Square frequently hosted political rallies and demonstrations. Hitler ordered the new headquarters renovated and re-christened as the Brown House after the colour of Nazi storm-trooper uniforms. The exterior was largely unchanged aside from some Nazi accoutrements, but the interior renovations were more substantive. The ground-floor vestibule became a flag hall displaying some of the oldest Nazi flags, including the party's most sacred relic: the blood-stained Blood Flag carried during the putsch. A grand staircase led to a second-floor foyer festooned with Nazi regalia. The centrepiece was the Senate Hall entranceway flanked by bronze plaques bearing the commemorative inscription later used on the Field Generals' Hall memorial. A bust of Dietrich Eckart, a co-founder of the Nazi Party, stood next to one of the plaques. The inclusion of Eckart conflated his death with the martyred putschists even though he died later of heart failure. This foreshadowed the impending conflation of party martyrdom to include the death of seemingly any German nationalist, First World War veterans, and eventually Second World War soldiers and civilians. Instead of functioning as the party's national headquarters, the Brown House basically evolved into a museum as most officials relocated to other buildings by 1937. This new repository of *Kampfzeit* commemoration mainly served as a reliquary for the Blood Flag and other objects embodying the memories of martyrdom.

Even before seizing power, Hitler ordered architect Paul Ludwig Troost, who had renovated the Brown House, to plan additional party buildings in the area. Troost eventually produced plans to redesign King's Square as an administrative, ceremonial and commemorative centre. Troost left the three nineteenth-century buildings surrounding the square unchanged (a Doric triumphal arch, a Corinthian gallery and an Ionic sculpture museum) but paved over the grassy interior to make the space better suited for marches and rallies. Troost designed two sets of matching buildings for the square's fourth side: two Temples of Honour sandwiched between two larger office buildings. The temples, square open-air atria standing about seven metres tall supported by modernist Doric columns, acquired intense meaning through Hitler's decision to re-inter the putsch martyrs there. The redesigned King's Square was heralded as the 'Forum of the Movement' providing 'a precinct of dignity and majesty consecrated through the remembrance of sacrifice, an Acropolis Germaniae' (Heilmeyer 1935: 140–141; Grammbitter 1995). Nazi propagandists also instrumentalized the temples to entrench notions of struggle, heroism, sacrifice and martyrdom into daily life.

The Nazi Party clearly desired visible, permanent markers locating its places of memory and narrating the memories of place. The regime also wanted this narrative performed in a literal sense; it had to be experienced by the masses through carefully choreographed re-enactments and ceremonies for dissemination to the widest possible audiences. Premiering in November 1933, the annual putsch commemorations became a highpoint in the new Nazi liturgy (Vondung 1971; Baird 1990; Hockerts 1993; Lehmbruch 1995; Pappert 2001; Köpf 2005; Heusler 2008). Hitler began the event with an evening speech at the Bürgerbräukeller. The following afternoon, Hitler, prominent Old Fighters and the Blood Flag retraced the original march route, passing by a plaque installed on the Ludwig Bridge before pausing for a moment of silence outside the Sterneckerbräu. The route was decorated profusely with flags, banners and several hundred tall pylons topped with firelit basins and wrapped in red fabric with the names of party martyrs emblazoned in gold. The martyrs' names were announced via loudspeaker as Hitler passed each pylon. Sombre drumbeats kept the procession's pace, while the Horst Wessel Anthem played in the background. The column halted at the Field Generals' Hall as a ceremonial cannon salvo thundered in the distance echoing the gunfire ten years earlier. Sixteen additional fire-topped pylons emblazoned with the martyrs' names lined the hall's main platform. After a moment of silence, Hitler laid a wreath before the memorial, largely concluding the ceremony.

The ceremony was relaunched in 1935 in its more or less final form. After Hitler's opening speech, 16 sarcophagi containing the martyrs' bodies were placed under an overnight honour guard at the Field Generals' Hall. Hitler and his followers marched along the same route, but now after performing their rituals at the hall, the procession continued towards King's Square accompanied by horse-drawn wagons transporting the sarcophagi. The column, accompanied by the national anthem, paused briefly in front of the Brown House before reaching the square, which was packed with formations of Nazi troopers except for a central pathway that allowed the procession to stride dramatically through the masses. Accompanied by the Blood Flag, Hitler and his entourage stood atop a reviewing platform as the rest of the procession formed up inside the square. Hitler's entourage then proceeded back across the square for the ceremonial climax, an honourary roll call of the martyred. The assembled masses shouted 'here!' in unison as the first name was read. The first sarcophagus was then placed inside one of the temples. The process was repeated until all 16 martyrs were re-interred and Hitler had placed wreaths before each sarcophagus.

The ceremony's symbolism was hard to miss. The first part of the march represented the struggles and sacrifices of the *Kampfzeit* culminating in martyrdom. The later portions of the march represented resurrection and victory as the martyrs assumed their posts as the Nazi movement's 'eternal guard'. The commemorations concluded at midnight as SS recruits assembled before the Field Generals' Hall to swear allegiance to Hitler and assume the role of martyrs (and murders)-in-waiting. Through this calculated use of places, processions and meticulously planned theatrics, the commemorations worked to condition participants and observers towards memories of struggle, heroism, sacrifice and martyrdom.

Local officials eagerly promoted Munich's claim as the heart of Nazi memory. A tourist guidebook commissioned by Mayor Karl Fiehler provides an evocative example. The inside cover directed readers in German, English, Spanish and Italian to a list of eight 'places sacred to the National Socialist Movement'. The list included the Temples of Honour, the Field Generals' Hall, the Brown House, the two new party office buildings adjacent the temples, the Bürgerbräukeller and Sternackerbräu breweries with the latter labelled 'the birthplace of the NSDAP', and finally the plaque marking Hitler's pre-war apartment (Spahn 1937: n.p.). These locations constituted an assemblage of memorial places and spaces scattered across Munich recounting Hitler's triumphal ascent from humble commoner to party leader. Munich was already a popular tourist destination, but the rise of the Nazi Party added a new set of sites/sights where the party faithful could pay homage or others could satisfy simple curiosity. As the celebratory tome edited by Fiehler explained:

> The mighty buildings of the Führer, the new reputation of Munich as the Capital of the Movement with its historical places from the *Kampfzeit*, and the announcement of the city as the City of German Art have led numerous domestic and foreign visitors here.
>
> (Fiehler 1937: 49)

The imperative to replicate these practices swept across Germany. Local party offices invariably featured a foyer or vestibule commemorating the Munich martyrs and other nationally prominent martyrs, like Horst Wessel or Albert Leo Schlageter. These spaces infused the mythology of martyrdom into daily bureaucracy while also providing backdrops for local commemorations, furthering the calculated interplay between symbolism, performance and experience. Artists created myriad paintings, photographs, postcards and other mementos propagating the mythology of 1923. The regime also circulated the ceremonies for those unable to visit personally through print, radio, newsreels and, most prominently, the 1936 film *For Us* (*Für Uns*). This propaganda freely intermingled images of the putsch sites with later commemorative activities, thereby blurring the distinction between past and present, participant and observer.

Commandeering German history

The Nazi Party was not content to commemorate, and in the process largely invent, its own history. The regime also endeavoured to recast all of German history as a story of racial war stretching back into prehistory. The Nazi Party's general historical narrative began with a distant golden age associated with either pagan Germanic tribes or the medieval Holy Roman Empire. Nazi ideologues touted the military accomplishments, real or imagined, of both periods but disagreed on which to emphasize. Pagan Germans lacked a strong central state, while medieval Germany remained highly fragmentary despite its imperial veneer and also adopted Christianity, interpreted as both foreign and Judaic. In both periods, Germans fought each other as often as their purported racial foes. Despite these debates, the Nazis were in broad agreement that the onset of rampant industrialization and urbanization during the nineteenth century had occasioned widespread racial degeneracy, miscegenation and decay that undermined traditional ways of life and threatened the survival of the German people. The Nazi Party promised salvation from these evils by uniting all ethnic Germans in a strong, centralized and expansionist state.

Building this narrative required a great deal of rhetorical and literary imagination, most evident in Hitler's rambling mediations in *Mein Kampf*. Hitler's subordinates had considerable room for initiative in this regard, so the regime's efforts to commandeer German history unfolded through disparate, ad hoc projects. For example, Braunschweig Minister President Dietrich Klagges instigated the renovation and eventual confiscation of the Braunschweig cathedral (Figure 14.2) as a 'state cathedral' honouring the twelfth-century Saxon Duke Heinrich the Lion, who was recast as a proto-Nazi racial warrior because of his campaigns against Slavic peoples. Klagges ordered a new crypt to house Heinrich's purported remains and emptied the church of its religious furnishings. The nave was decorated with a series of murals depicting Heinrich's campaigns, and a massive eagle clutching a wreathed swastika soaring above rows of swastika flags replaced the main altar. The result of these renovations, one local historian concluded, was that 'the cathedral of Heinrich the Lion is now more than an artistic, medieval church building, it is a holy place of the German nation' (Flechsig 1939: 365; Fuhrmeister 2009). Other Nazi leaders focused on prehistory. Franconian boss Julius Streicher, for example, aggressively promoted the Hesselberg mountain as an important pagan shrine that should again be revered as a new 'site of pilgrimage' (Greif 2007: 333). Streicher staged massive rallies that drew tens of thousands to the summit and eventually included plans for a new Adolf Hitler School and various other party facilities, although little progress was made.

There were efforts to organize more systematic programmes. Bavarian President Ludwig Siebert launched a major restoration campaign encompassing dozens of medieval castles, ruins and city fortifications. Siebert's programme focused on

DER BRAUN-
SCHWEIGISCHE
STAATSDOM
MIT DER GRUFT
HEINRICHS
DES LÖWEN

EIN VORBILD
GEGENWARTSNAHER
DENKMALSPFLEGE
IM NEUEN DEUTSCHLAND

Figure 14.2 Braunschweig cathedral.

Source: Flechsig (1939: 358).

purging these structures of later additions in an attempt to return them to some 'pure' state and then making them available for party functions and public visits. The Trifels Castle in the Rhineland was much more ambitious. The castle was basically a pile of rubble, so Siebert ordered its complete reconstruction as a 'national holy place' that would become a 'symbol of the inner connection of the new Reich with the old and thereby a symbol of the immortality of the German spirit' (Siebert 1941: 9). Work proceeded rapidly until interrupted by war. Despite their differences, Siebert, Streicher and Klagges shared a belief that their quasi-sacred places of memory constituted a vital resource in shaping contemporary perceptions of the past. All they had to do was create the appropriate atmosphere allowing visitors to these locations to experience history from a Nazi perspective.

SS chief Heinrich Himmler acted similiarily by confiscating the Quedlinburg abbey church after his team of archaeologists discovered the alleged remains of

Heinrich I, commonly regarded as Germany's first king in the early tenth century. Himmler admired Heinrich for his military campaigns against the Slavs and Magyars and ordered a new crypt and sarcophagus. Himmler was also interested in Widukind, an eighth-century noble who resisted Charlemagne's attempts to subjugate and convert the Saxons. Himmler sponsored commemorations at Widukind's tomb in the town of Enger and converted a nearby medieval merchant home into a Widukind memorial that hosted various SS ceremonies. These commemorative places reflected Himmler's idiosyncrasies, and, in any event, they were limited to exclusive use by the SS. Himmler also sponsored archaeological digs to collect relics dating to Germanic prehistory. SS researchers used these finds to advance their racially infused understanding of history while also providing the basis for the construction of open-air museums dedicated to Germanic prehistory, most notably the re-created Germanic Farmstead opened near Oerlinghausen in 1936 (Schmidt 2002; Kaldewei 2006; Moors 2009). Most of these efforts proved short-lived as Himmler and the SS were soon fixated on war and genocide. Himmler's exact intentions regarding these memorial projects remain a matter of debate and often wild speculation. At a minimum, it seems clear that Himmler intended to carve out a distinct SS identity within the Nazi movement predicated on the existence of an elite class of racial warriors who had protected the German people through history. Himmler sought to expropriate places he believed would help cultivate this ethos among his subordinates and ultimately condition them for unimaginable brutality against anybody who ran afoul of SS interests.

These projects generally relied on existing structures, but the Stedings' Honour (*Stedingsehre*) amphitheatre located near Bremen was an attempt to create a new place of memory. In 1234, the church ordered a crusade against the local Stedinger farmers, who were massacred. The conflict was largely forgotten until regional party boss Carl Röver commissioned a new amphitheatre for dramatic re-enactments marking the conflict's 700th anniversary. The complex featured a mock peasant village of thatched-roofed cottages separated by a moat from a seating area for approximately 7,800 spectators. Leading Nazis including Himmler, Alfred Rosenberg and Richard Walther Darré attended the foundation-stone ceremony in October 1934. The first performances followed in 1935 before sold out audiences. The play portrayed the Stedingers as virtuous peasants valiantly defending their homes and their beliefs to the death. Röver soon developed plans to augment this 'Cult Site Stedingen' with new schools and other training facilities (Finsterhölzl 1999). War prevented the expansion, but the theatre nonetheless contributed to the regime's glorification of struggle, heroism, sacrifice and martyrdom. Darré, Himmler and Rosenberg also collaborated to create the Saxon Grove (*Sachsenhain*) memorial commemorating Germans fighting and dying defending their beliefs, in this case at the Massacre of Verden where Charlemagne executed around 4,500 Saxons for refusing conversion to Christianity. Designers relocated old farmsteads to recreate a medieval hamlet in a clump of forest purported to be the execution site. Footpaths lined with 4,500 boulders leading to the hamlet were largely completed by 1937 (Kaldewei 2006). Like Stedings' Honor, the Saxon Grove was reimagined as a place where Germanic forebears defended their beliefs and community to the death against foreign invaders.

These cases are illustrative of the regime's determination to reinterpret Germany history through Nazi ideology. But beyond simply rewriting the history books, which they also did, Nazi officials sought to redesign specific places believed to represent their revisionist narratives. The ubiquitous references to religion and pilgrimage made clear that the German masses were expected to experience these places personally. The act of journeying to these locations carried significance that was assumed to amplify the reverential affect of the actual site/sight visit. The experience of visiting these places of memory would in turn cultivate memories of these places among the general public. The ultimate goal was the transformation of German history into a narrative of racial war precursory to salvation through submission to Nazi ideology.

Remembering war

The First World War was a seminal event in German history and for many top Nazis who were veterans. Predictably, the war loomed large in Nazi rhetoric and commemorations. As a fledgling movement, the party had little choice but to rely initially on existing monuments, memorials and public spaces. The Hall of Honor (*Ehrenhalle*) in Nuremberg was one of the most prominent. Completed in 1930, the hall commemorated local soldiers who died in the First World War, but the Nazis, who used the surrounding parkland for rallies, quickly co-opted the memorial. The hall was a rather austere, colonnaded archway with a smooth stone exterior facing a paved patio flanked by square pylons. Soon after taking power, Hitler ordered the surrounding park transformed into a grand arena for the Nuremberg rallies. The hall became a focal point opposite Hitler's speaking podium flanked by massive grandstands. Hitler provided a high point of the rallies as he marched from the podium to the hall along a granite pathway that cut through massed formations of Nazi troopers. Upon reaching the hall, Hitler commemorated the war dead and the Munich martyrs, thereby conflating Germany's war dead with those who later died for the party. Indeed, Hitler soon had military representatives participating in party commemorations, most prominently Commander in Chief of the Armed Forces General Werner von Blomberg laid a wreath at the Field Generals' Hall during the 1935 putsch ceremony, further blurring the distinction between the military dead and party martyrs.

The Nazi Party had little trouble incorporating existing monuments into their public rituals since most already reflected a rather pompous, chauvinistic nationalism. Memorials expressing anti-war messages or feelings of regret or grief, such as the New Guardhouse (*Neue Wache*) in Berlin, were more problematic but usually redesigned with relative ease. One of the regime's most prominent interventions involved the Tannenberg Memorial commemorating Germany's victory under General Paul von Hindenburg against Russian forces during the First World War. A private campaign organized by veteran groups led to the memorial's dedication in 1927. The memorial took the form of a stone octagon with a square tower centred along each side. The interior was an open grassy space with four walkways converging on a central platform topped by a bronze cross. Bodies of unknown German war dead were buried below. Hitler ordered the memorial redesigned after Hindenburg's death in 1934. The interior was paved and a new crypt for Hindenburg and the unknown soldiers guarded by oversized stone soldiers was added across from the main entrance. Troost's widow, writing in a popular pictorial overview of Nazi architecture, excitedly proclaimed the Tannenberg Memorial to be a 'great Castle of the Dead (*Totenburg*)' (Troost 1938: 34; Tietz 1999).

Local officials were eager to replicate these higher-profile commemorative practices in their communities. Renaming places offered the quickest means, so Germany promptly sported myriad Adolf Hitler Squares, Horst Wessel Streets and similar places (Baird 1990). Hitler worried the rapid proliferation trivialized the party and ordered that honorary and commemorative names be limited to new places. The streets of a new suburb in Regensburg were named after cities and territories that Germany lost after the war (i.e. Danzig) or had sizeable ethnic German populations (i.e. Sudetenland), for example, while a new suburb in Munich's bore the names of the 16 putsch martyrs. Monuments to the party were also erected in countless communities. Some commemorated specific figures such as Wessel, while others were more generic referents to the party. The monuments were rather unremarkable, often following conventional forms such as an obelisk, and the awarding of honorary names is hardly unique to Nazism. Yet the speed at which new names and markers appeared was striking. The small Franconian town of Gunzenhausen dedicated what was possibly the first party monument, a relatively simple stone staircase leading up to a 7.5 metre obelisk, in April 1933 (Greif 2007: 331–332). The overall effect was to saturate public spaces with the sense of a movement in a constant struggle for national survival. Hitler and his lieutenants used this atmosphere to justify calls for ever greater obedience and sacrifice from the people.

Hitler showed little personal interest in these commemorative practices outside of Munich, Nuremberg and, of course, Berlin. Hitler envisioned a thorough restructuring of Berlin around broad boulevards, massive governmental buildings and overpowering monuments. Albert Speer, Hitler's chief architect, developed plans for a long north–south boulevard linking a new railway station with a gigantic government centre. A massive triumphal arch would dominate the middle of the boulevard. Standing about 120 metres high, the arch would dwarf Paris' Arc de Triomphe and provide enough room to bear the names of Germany's 1.8 million First World War dead. A massive concrete cylinder was poured to test whether the ground could support such a massive structure, but the project otherwise stayed on the drawing board. Plans for Munich's centrepiece monument did not make it that far. It was originally intended as a memorial to Munich's 'liberation' from leftist forces in 1919, but Hitler ordered it changed to a 'Monument of the Movement' by 1940. Various designs were produced by 1944, but the basic idea was a massive obelisk standing approximately 212 metres tall. An eagle with outstretched wings clutching a wreathed swastika in its claws topped the obelisk (Rasp 1981: 201; Nerdinger 1993: 353–354; Weihsmann 1998: 272–278).

The monument's evolution was another example of the party gradually subsuming ever more of Germany's history and dead into its liturgical pantheon. Both monuments had explicit representational functions, but they were also entangled with geographies of performance and emotion. Set amid expansive boulevards lined with blocks of immense buildings, the monuments were to frame a broader stage where displays of regime power and public adulation could be performed on a colossal scale. Hitler demanded these new showpiece districts assume gargantuan proportions to awe, inspire and overwhelm even casual visitors and ultimately reach specific conclusions about the political prowess of the Nazi's thousand-year Reich.

The regime also sought to commemorate Second World War dead through massive monuments. In 1941, Wilhelm Kreis, a respected and aged architect, was tasked with designing memorials for the ongoing conflict (Figure 14.3). Kreis proposed a series of massive cenotaphs marking the locations of major battlefields stretching 'from Narvik to Africa, from the Atlantic to the plains of Russia, beginning at the borders of the Greater German Reich and ending on the front lines of the greatest battle in the world' (Tamms 1943: 51). The monuments would be positioned in highly visible locations and draw inspiration from local materials and architecture. The North African monument, for example, resembled an ancient Egyptian mastaba. The largest monument would tower 150 metres over the

Figure 14.3 Kreis designed several massive war memorials, including this one planned on conquered Russian territory.

Source: Tamms (1943: 52).

Note: All three figure sources were published by Franz Eher Verlag, which bore the title Central Publisher (*Zentralverlag*) of the NSDAP and was outlawed in 1945 due to its Nazi connections.

Russian steppe with a massive interior crypt reaching 100 metres tall. Kreis' plans drew from the ongoing efforts of the German War Graves Commission, a private organization that supervised the country's First World War cemeteries across Europe. The cemeteries were normally centred on monumental stone cenotaphs placed on hilltops or other prominent natural settings. Like the Tannenberg Memorial, the cenotaphs invited visitors into a crypt-like space, commonly sunken into the ground to accentuate a grave-like sense of place so that, as one author put it, visitors journeyed 'into the deep, the mysterious, harrowing impression of a Romanesque crypt, thereby renewing the connection of the warriors' cenotaph with the neighboring graves' (Gstettner 1940: 154).

Kreis also designed the Soldiers' Hall, a heavy-set, neoclassicist structure to be built along Speer's grand boulevard in Berlin. The structure, described as 'a new, true German cathedral', featured an airy, vaulted interior bathed in light from tall entranceways. A cavernous, darkened crypt rested below. In addition to its commemorative function, the hall spoke 'more of the vows of the living as of the legacy of the dead' (Tamms 1943: 57), so that its primary function was to precondition the living to sacrifice their lives for the regime instead of commemorating the fallen. Troost's widow enthused that Kreis' memorials and monuments meant that 'the meaning of struggle, sacrifice and victory will find in them eternal form in lasting stone' (Troost 1943: 7), but in many ways, these and the Nazi Party's other commemorative places and practices amounted to a hideous cult of death that mutated the remembrance of those who had died for Germany into performances of past and future martyrdom on behalf of the party.

By the time the Second World War reached its bloody midpoint, the regime had fully subsumed all German war dead into the pantheon of Nazi martyrdom. Party guidelines issued in 1942 began succinctly by stating:

> The 9th November unites the entire national community in remembrance of the dead of the movement, but also in remembrance of the fallen of the First World War and the current war. . . . The 16 dead of the Field Generals' Hall and all blood victims of the movement are therefore joined in a firm connectedness with the fallen heroes of this war.
>
> (Anon. 1942: 492)

The guidelines followed this declaration with a ceremonial programme, replete with its own hymns, readings, vows and oaths dedicated to the 'holy dead' (Anon. 1942: 498). It was equally important that the venue be solemn with decorations limited to party flags and symbols. Caskets and other funerary trappings were expressly forbidden. The overall result was the creation of ceremonial spaces focused on glorifying unwavering sacrifice and martyrdom instead of remembrance and mourning.

Concluding thoughts

The Nazi Party has long been likened to a political religion (Vondung 1971; Burleigh 2000), but the movement defies easy generalizations. Soon after seizing power, the Nazi regime instigated a broad-based reconfiguration of places of memory and memories of places across Germany. These commemorative practices clearly drew from established religious rhetoric and ritual in an effort to celebrate struggle, heroism, sacrifice and death on behalf of the movement. One primer for the Hitler Youth captured this chilling sentiment by noting Germany's war dead were interred in 'sites of dying, sites of belief, holy places of the Reich' (Dörner 1939: 156). The author enumerated the main fronts of the First World War before drawing a direct connection to the party martyrs:

> In reverence and silent readiness, we remember our dead – the dead of the war and the dead of our movement. . . . The names of the dead of our movement reverberate across the land, the names of the battlefield cemeteries that form the sacred ring around the Reich resonate in remembrance. In us, there is no grief, only obedience and readiness reign.
>
> (Dörner 1939: 156)

The regime's cynical perspectives on places of memory and martyrdom positioned the Nazi movement as the culmination of a sweeping historical ontology of racial struggle and war. These diffuse sites/sights of memory anchored that narrative in physical space and place. The regime also demanded active participation in its commemorative liturgy of rallies, marches and re-enactments so that individuals acquired tangible experiences of belonging to the Nazi racial community. Nazi pageantry embraced these carefully staged combinations of representation, action and experience to initiate an ever-growing number of Germans into the pantheon of party martyrs. In that sense, these places of Nazi memory and memories of Nazi places served as real and imagined vehicles for stitching together semiotics, performativity and affect to (re)produce emergent geographies of National Socialism.

References

Adey, P. 2008. Airports, mobility and the calculative architecture of affective control. *Geoforum* 39 (1): 438–451.

Alderman, D. and Inwood, J. 2013. Landscapes of memory and socially just futures. In N. Johnson, R. Schein and J. Winders (eds) *The Wiley-Blackwell Companion to Cultural Geography*. Chichester: Wiley-Blackwell, 186–197.

Allen, J. 2006. Ambient power: Berlin's Potsdamer Platz and the seductive logic of public spaces. *Urban Studies* 43 (2): 441–455.

Anderson, B. and Harrison, P. (eds) 2010. *Taking-place: Non-representational Theories and Geography*. Farnham: Ashgate.

Anon. 1942. Zum 9 November 1942: Gedenktag für die Gefallen der Bewegung. *Die neue Gemeinschaft* 8 (9): 492–502.

Baird, J. 1990. *To Die for Germany: Heroes in the Nazi Pantheon*. Bloomington: Indiana University Press.

Bartetzko, D. 1985. *Zwischen Zucht und Ekstase: Zur Theatralik von NS-Architektur*. Berlin: Mann.

Bauer, R., Hockerts, H.G., Schütz, B., Till, W. and Ziegler, W. (eds) 1993. *München: 'Haupstadt der Bewegung': Bayerns Metropole und der Nationalsozialismus*. Munich: Klinkhardt und Biermann.

Behrenbeck, S. 1996. *Der Kult um die Toten Helden: Nationalsozialistische Mythen, Riten und Symbole 1923 bis 1945*. Vierow: SH-Verlag.

Burleigh, M. 2000. National Socialism as a political religion. *Totalitarian Movements and Political Religions* 1 (2): 1–26.

Curti, G.H. 2008. From a wall of bodies to a body of walls: politics of affect/politics of memory/politics of war. *Emotion, Space and Society* 1 (2): 106–118.

Dörner, C. 1939. *Das Deutsche Jahr: Feiern der Jungen Nation*. Munich: Franz Eher.

Dovey, K. 2009. *Becoming Places: Urbanism/Architecture/Identity/Power*. London: Routledge.

Dresler, A. 1939. *Das Braune Haus und die Verwaltungsgebäude der Reichsleitung der NSDAP*. Munich: Franz Eher.

Espe, W.M. 1933. *Das Buch der N.S.D.A.P.: Werden, Kampf und Ziel*. Berlin: G. Schönfeld.

Fiehler, K. (ed.) 1937. *München baut auf: Ein Tatsachen- und Bildbericht über den Nationalsozialistischen Aufbau in der Hauptstadt der Bewegung*. Munich: Franz Eher.

Finsterhölzl, C. 1999. Die Einweihung der Niederdeutschen Gedenkstätten 'Stedingsehre': Ein Beispiel der nationalsozialistischer Selbstinszenierung im Gau Weser-Ems. *Oldenburger Jahrbuch* 99: 177–205.

Flechsig, W. 1939. Der braunschweigische Staatsdom mit der Gruft Heinrichs des Löwen. *Die Kunst im deutschen Reich, Ausgabe B* 3 (11): 358–365.

Foote, K.E. and Azaryahu, M. 2007. Toward a geography of memory: geographical dimensions of public memory and commemoration. *Journal of Political & Military Sociology* 35 (1): 125–144.

François, E. and Schulze, H. (eds) 2001. (eds) *Deutsche Erinnerungsorte*. Munich: Beck.

Fuhrmeister, C. 2009. Purifizierung, Moderne, Ideologie: Zur Umgestaltung des Braunschweiger Doms im Nationalsozialismus. In D. Rammler and M. Strauß (eds) *Kirchenbau im Nationalsozialismus: Beispiele aus der braunschweigischen Landeskirche*. Wolfenbüttel: Evangelische-lutherische Landeskirche in Braunschweig.

Grammbitter, U. 1995. Vom Parteiheim in der Briener Straße zu den Monumentalbauten am 'Königlichen Platz'. In I. Lauterbach, J. Rosenfeldt and P. Steinle (eds) *Bürokratie und Kult: Das Parteizentrum der NSDAP am Königsplatz in München*. Munich: Deutscher Kunstverlag, 61–87.

Greif, T. 2007. *Frankens Braune Wallfahrt: Der Hesselberg im Dritten Reich*. Ansbach: Historischen Vereins für Mittelfranken.

Gstettner, H. 1940. Die deutsche Gestaltung des Kriegergrabes. *Kunst für Alle* 55 (7): 144–154.

Heilmeyer, A. 1935. Die Stadt Adolf Hitlers. *Süddeutsche Monatshefte* 33: 135–141.

Heusler, A. 2008. *Das braune Haus: Wie München zur 'Hauptstadt der Bewegung' wurde*. Munich: Deutsche Verlags-Anstalt.

Hockerts, H.G. 1993. Mythos, Kult und Feste: München im nationalsozialistischen 'Feierjahr'. In R. Bauer, H.G. Hockerts, B. Schütz, W.Till and W. Ziegler (eds) *München: 'Haupstadt der Bewegung': Bayerns Metropole und der Nationalsozialismus*. Munich: Klinkhardt und Biermann, 331–341.

Hoelscher, S. and Alderman, D. 2004. Memory and place: geographies of a critical relationship. *Social & Cultural Geography* 5 (3): 347–355.

Horton, J. and Kraftl, P. 2014. *Cultural Geographies: An Introduction*. London: Routledge.

Jacobs, J. and Merriman, P. 2011. Practising architectures. *Social & Cultural Geography* 12 (3): 211–222.

Jones, O. 2011. Geography, memory and non-representational geographies. *Geography Compass* 5 (12): 875–885.

Jones, O. and Garde-Hansen, J. (eds) 2012a. *Geography and Memory: Explorations in Identity, Place and Becoming*. Basingstoke: Palgrave Macmillan.

Jones, O. and Garde-Hansen, J. 2012b. Introduction. In O. Jones and J. Garde-Hansen (eds) *Geography and Memory: Explorations in Identity, Place and Becoming*. Basingstoke: Palgrave Macmillan, 1–23.

Kaldewei, G. 2006. *'Stedlingsehre' soll für ganz Deutschland ein Wallfahrtsort werden: Dokumentation und Geschichte einer NS-Kultstätte auf dem Bookholzberg 1934–2005*. Delmenhorst: Aschenbeck & Holstein.

Kershaw, I. 1993. 'Working towards the Führer': reflections on the nature of the Hitler dictatorship. *Contemporary European History* 2 (2): 103–118.

Köpf, P. 2005. *Der Königsplatz in München: Ein deutscher Ort*. Berlin: Ch. Links.

Koshar, R. 1998. *Germany's Transient Pasts: Preservation and National Memory in the Twentieth Century*. Chapel Hill: University of North Carolina Press.

Kraftl, P. and Adey, P. 2008. Architecture/affect/inhabitation: geographies of being-in buildings. *Annals of the Association of American Geographers* 98 (1): 213–231.

Kriegl, H. 2003. *Adolf Hitlers 'treueste Stadt' Landsberg am Lech 1933–1945*. Nuremberg: W. Tümmels.

Lees, L. 2001. Towards a critical geography of architecture: the case of an ersatz Colosseum. *Ecumene* 8 (1): 51–86.

Lees, L. and Baxter, R. 2011. A 'building event' of fear: thinking through the geography of architecture. *Social & Cultural Geography* 12 (2): 107–122.

Lehmbruch, H. 1995. Acropolis Germaniae: Der Königsplatz – Forum der NSDAP. In I. Lauterbach, J. Rosenfeldt and P. Steinle (eds) *Bürokratie und Kult: Das Parteizentrum der NSDAP am Königsplatz in München*. Munich: Deutscher Kunstverlag, 17–45.

Liese, H. (ed., on behalf of the Hauptkulturamt in der Reichspropagandaleitung der NSDAP). 1943. *Ich kämpfe*. Munich: Franz Eher.

Lorimer, H. 2005. Cultural geography: the busyness of being 'more-than-representational'. *Progress in Human Geography* 29 (1): 83–94.

McCormack, D. 2008. Engineering affective atmospheres: on the moving geographies of the 1897 Andree expedition. *Cultural Geographies* 15 (4): 413–430.

Macdonald, S. 2006. Word in stone? Agency and identity in a Nazi landscape. *Journal of Material Culture* 11 (1/2): 105–126.

Maier Hartmann, F. 1942. *Die Bauten der NSDAP in der Hauptstadt der Bewegung*. Munich: Franz Eher.

Meusburger, P., Heffernan, M. and Wunder, E. (eds) 2011. *Cultural Memories: The Geographical Point of View*. Dordrecht: Springer.

Moors, M. 2009. Das 'Reichshaus der SS-Gruppenführer': Himmler Pläne und Absichten in Wewelsburg. In J.E. Schulte (ed.) *Die SS, Himmler und die Wewelsburg*. Paderborn: Ferdinand Schöningh, 161–179.

Mosse, G.L. 1990. *Fallen Soldiers: Reshaping the Memory of the World Wars*. New York: Oxford University Press.

Nerdinger, W. (ed.) 1993. *Bauen im Nationalsozialismus: Bayern 1933–1945*. Munich: Architekturmuseum der Technischen Universität München.

Pappert, L. 2001. *Der Hitlerputsch und seine Mythologisierung im Dritten Reich*. Neuried: ars una.

Rasp, H. 1981. *Eine Stadt für tausend Jahre: München – Bauten und Projekte für die Hauptstadt der Bewegung.* Munich: Süddeutscher.

Reich, A. and Achenbach, O.R. 1933. *Vom 9. November 1918 zum 9. November 1923: Die Entstehung der deutschen Freiheitsbewegung.* Munich: Franz Eher.

Rose, G., Degen, M. and Badas, B. 2010. More on 'big things': building events and feelings. *Transactions of the Institute of British Geographers* 35 (3): 334–349.

Sakamoto, R. 2015. Mobilizing affect for collective war memory. *Cultural Studies* 29 (2): 158–184.

Schmidt, M. 2002. Die Rolle der musealen Vermittlung in der nationalsozialistischen Bildungspolitik: Die Freilichtmuseen deutscher Vorzeit am Beispiel von Oerlinghausen. In A. Leube, with M. Hegewisch (eds) *Prähistorie und Nationalsozialismus: Die mittel- und osteuropäische Ur- und Frühgeschichtsforschung in den Jahren 1933–1945.* Heidelberg: Synchron, 147–159.

Schramm, K. 2011. Landscapes of violence: memory and sacred space. *History & Memory* 23 (1): 6–22.

Schüßler, R. 1935. *Das Sternecker-Museum der NSDAP in München, Tal 54.* Munich: R. Schüßler.

Schumacher, R. von. 1935. *Deutschland-Fibel: Volk-Raum-Staat.* Second edn. Berlin: Offene Worte.

Siebert, L. 1941. Deutsches Kulturschaffen als völkische Pflicht. In L. Siebert (ed.) *Wiedererstandene Baudenkmäle: Ausgewählte Arbeiten aus dem Ludwig-Siebert-Programm zur Erhaltung bayerische Baudenkmäle.* Munich: F. Bruckmann, 7–10.

Simonsen, K. 2007. Practice, spatiality and embodied emotions: an outline of a geography of practice. *Human Affairs* 17 (2): 168–181.

Spahn, B. (ed.) 1937. *München: Hauptstadt der Bewegung.* Berlin: Riegler.

Tamms, F. 1943. Die Kriegerehrenmäler von Wilhelm Kreis. *Die Kunst im Deutschen Reich,, Ausgabe B* 7 (3): 50–57.

Taylor, B. 2007. *Hitler's Headquarters: From Beer Hall to Bunker, 1920–1945.* Washington, DC: Potomac.

Thrift, N. 2008. *Non-Representational Theory: Space/Politics/Affect.* London: Routledge.

Tietz, J. 1999. *Tannenberg-Nationaldenkmal.* Berlin: Bauwesen.

Troost, G. 1938. *Das Bauen im neuen Reich.* First edn. Bayreuth: Gauverlag Bayerische Ostmark.

Troost, G. 1943. *Das Bauen im neuen Reich.* Second edn. Bayreuth: Gauverlag Bayerische Ostmark.

Tyner, J., Inwood, J. and Alderman, D. 2014. Theorizing violence and the dialectics of landscape memorialization: a case study of Greensboro, North Carolina. *Environment and Planning D: Society and Space* 32 (5): 902–914.

Vondung, K. 1971. *Magie und Manipulation: Ideologischer Kult und politische Religion des Nationalsozialismus.* Göttingen: Vandenhoeck & Ruprecht.

Waterton, E. and Watson, S. 2013. Framing theory: towards a critical imagination in heritage studies. *International Journal of Heritage Studies* 19 (6): 546–561.

Weihsmann, H. 1998. *Bauen unterm Hakenkreuz: Architektur des Untergangs.* Vienna: Promedia.

Whatmore, S. 2006. Materialist returns: practising cultural geography in and for a more-than-human world. *Cultural Geographies* 13 (4): 600–609.

Index

Page numbers in *italics* denotes an illustration